A History of Fruit Varieties

The American Pomological Society
One Hundred Fifty Years
1848 - 1998

A History of Fruit Varieties

The American Pomological Society
One Hundred Fifty Years
1848 - 1998

EDITED BY
David C. Ferree, Ph.D.,
American Pomological Society

ILLUSTRATED BY
Lynda Eades Chandler

DESIGNED AND PUBLISHED BY
Good Fruit Grower Magazine
Yakima, Washington

This book resulted from a collaboration between the American Pomological Society
and *Good Fruit Grower* magazine, nonprofit organizations
serving fruit growers throughout the world.

©1998 by *Good Fruit Grower* Magazine

All rights reserved

Printed in the United States of America

First published 1998

The paper in this publication meets the minimum requirements
of the American National Standard for Information Sciences—
Permanence of Paper for Printed Library Materials, ANSI Z39.48-1984

Library of Congress Cataloging-in-Publication Data

A history of fruit varieties : the American Pomological Society : one hundred fifty years, 1848-1998 / edited by David C. Ferree ; illustrated by Lynda E. Chandler.
    p.   cm.

Articles originally appeared in the *Fruit Varieties Journal*.

Includes bibliographical references and index.

ISBN 0-9630659-7-1 (alk. paper)

1. Fruit--Varieties--History. 2. Nuts--Varieties--History. 3. Botanical illustration. I. Ferree, David C. (David Curtis), 1943- . II. Chandler, Lynda E. III. American Pomological Society. IV. *Good Fruit Grower* magazine.

SB357.33.H57 1998

634'047--dc21
                                                                                     98-34446
                                                                                              CIP

Preface

To commemorate 150 years of service to fruit growers and to acknowledge the many professionals who have loyally served this industry, the American Pomological Society has chosen to republish the last 12 years of cover stories from its quarterly, the *Fruit Varieties Journal*. This publication, *A History of Fruit Varieties*, provides, in one place, the history and impact these fruit cultivars have made.

Botanical illustrations for each of these cultivars were produced by artist Lynda Chandler and are featured with each article. In addition, Ms. Chandler designed the full-color cover artwork.

Appreciation is extended to authors for the time and effort required to piece together the historical information for each article. The help of Jim Black and others at *Good Fruit Grower* magazine was invaluable in making the book a reality.

It is my hope that the American Pomological Society will continue its long tradition of helping the fruit industry, and that this book, in a small way, recognizes the contributions of the Society and its membership.

David C. Ferree, Professor
Department of Horticulture and Crop Science
Ohio State University
Wooster, Ohio

Table of Contents

American Pomological Society, 150 Year History ix

The 'Delicious' Apple
by Carlos Fear & Paul Domoto 1

The 'Bartlett' Pear
by Craig Chandler 5

The 'Valencia' Orange
by R.K. Soost 7

The 'Concord' Grape
by C.A. Cahoon 9

The 'Searles' Cranberry
by Elden Stang 12

The 'Redhaven' Peach
by A.F. Iezzoni 15

The 'Napoleon' Sweet Cherry
by Susan Brown 17

The 'Golden Delicious' Apple
by Tara Auxt Baugher & Steve Blizzard 19

The 'Kiwifruit'
by I.J. Warrington 22

The 'Jonathan' Apple
by Roy Rom 25

The 'Montmorency' Sour Cherry
by Amy Iezzoni 30

The 'Stuart' Pecan
by Tommy Thompson 33

The 'York Imperial' Apple
by Howard Rollins, Jr. 36

The 'Willamette' Red Raspberry
by Hugh Daubeny, F.J. Lawrence, & G.R. McGregor 38

The 'Barcelona' Hazelnut
by Shawn Mehlenbacher & Anita Miller 41

The 'Elberta' Peach
by Stephen Myers, W.R. Okie, & Gary Lightner 47

The 'Bluecrop' Highbush Blueberry
by Arlen Draper & Jim Hancock 55

The 'McIntosh' Apple
by John Proctor 57

The 'Bing' Sweet Cherry
by Teryl Roper & Curt Rom 60

The Australian Pistachio 'Sirora'
by D.H. Maggs 63

The 'Gala' Apple
by A.G. White 65

The 'Jonagold' Apple
by Roger Way & Susan Brown 67

The 'Earliglow' Strawberry
by Gene Galletta & John Maas 72

The 'Springcrest' Peach
by W.R. Okie & S.C. Myers 75

The 'Heritage' Red Raspberry
by Hugh Daubeny, Kevin Maloney, & G. McGregor 78

The 'Flordaprince' Peach
by W.B. Sherman & P.M. Lyrene 81

The 'Wealthy' Apple
by Teryl Roper 83

The 'Sharpblue' Southern Highbush Blueberry
by P.M. Lyrene & W.B. Sherman 86

The 'Meeker' Red Raspberry
by Patrick Moore & Hugh Daubeny 90

The 'Empire' Apple
    by M. Derkacz, D.C. Elfving, & C.G. Forshey 94

The 'Marsh' Grapefruit
    by Federick Gmitter, Jr. 97

The 'Totem' Strawberry
    by Hugh Daubeny, F.J. Lawrence, & P.P. Moore 102

The Lingonberry
    by Elden Stang. 106

The 'Granny Smith' Apple
    by Ian Warrington. 111

The 'Reyna' ('Alfajayucan') Cactus Pear
    by Candelario Mondragón Jacobo
    & Salvador Pérez González 116

The 'Schley' Pecan
    by Darrell Sparks. 119

The 'Glen Moy' Red Raspberry
    by Derek Jennings 125

The 'Western Schley' Pecan
    by Darrell Sparks. 128

The 'Senga Sengana' Strawberry
    by Edward Zurawicz & Hugh Daubeny 133

The 'Fuji' Apple
    by Yoshio Yoshida, Xuetong Fan, & Max Patterson . 137

The 'Sumner' Pecan
    by Darrell Sparks. 142

The 'Crandall' Black Currant
    by Kim Hummer & Stanislaw Pluta 146

The 'Kent' Strawberry
    by Andrew Jamieson 149

French Hybrid Grapes in North America
    by G.A. Cahoon. 152

The 'Desirable' Pecan
    by Darrell Sparks. 169

The 'Cortland' Apple
    by Roger Way & Susan Brown. 175

The 'Marion' Trailing Blackberry
    by Chad Finn, Bernadine Strik, &
    Francis Lawrence. 183

The 'Sunred' Nectarine
    by Jeffrey Williamson & Wayne Sherman 188

Wilder Awards 191
Index 195

The American Pomological Society: One Hundred Fifty Years

*by R.M. Crassweller, Professor of Tree Fruit,
Department of Horticulture, Penn State University*

In 1998, we celebrate the 150th anniversary of the American Pomological Society. The Society is a worldwide organization with members located on six continents. The *Fruit Varieties Journal* published by the Society has become a repository for information on fruit cultivar performance and new releases. The Society continues to meet on an annual basis and sponsors periodic symposia at various scientific gatherings on the subject of fruit and nut cultivar performance. Over the intervening years since its founding in 1848, the Society has changed and evolved into its present form. The change and evolution was the result of the dedication of the founding members, their descendants, and our current membership.

The Society had its inception at a September 1848 meeting in Buffalo, New York. Some of the leading pomologists of the time, John Warder, Patrick Barry, and A.J. Downing, under the auspices of the New York Agricultural Society, convened the first national assembly solely for the discussion of pomological subjects. The purposes of the original meeting were, "...apart from discussion, to identify synonyms, to correct errors in the names of our fruit, and to establish a uniform nomenclature."

Twelve states were represented at the session, which lasted three days. At the end of this meeting, the assembly resolved to perpetuate itself by holding an annual meeting under the name of the North American Pomological Convention.

About one month later, another group of more eastern fruit growers organized under the name of National Congress of Fruit Growers. They met in New York City under the sponsorship of the American Institute as a result of a call from the horticultural societies of Massachusetts, Pennsylvania, and New Jersey. Among the members at this meeting was Marshall Pinckney Wilder, who presided as president. The objectives proposed for this convention were:

To compare fruits from various sources and localities, with a view of arriving at correct conclusions as to their merits, and to settle doubtful points respecting them.
To assist in determining the synonyms by which the same fruit is known in different parts of the country.
To compare opinions respecting the value of the numerous varieties already in cultivation, and to endeavor to abridge, by general consent, the long catalogue of indifferent or worthless sorts at the present time propagated by nurserymen and fruit growers.
To elicit and disseminate pomological information, and to maintain a cordial spirit of intercourse among horticulturists.

The following year, representatives from the two societies assembled in Syracuse, New York, on September 14 and agreed to unite in one society at the American Pomological Convention. The united societies met together for the first time in the autumn of 1850 in Cincinnati, Ohio. Dr. W.D. Brincklé of Philadelphia was selected as president and held that position for the first two years of the Society's existence. The need for the adoption of a constitution and bylaws was not addressed until the next meeting of the Society was held in 1852 in Philadelphia.

At the meeting in Philadelphia, a business committee proposed the following duties for the newly formed Society:

To revise the lists of fruits recommended by the former session of the Congress for general cultivation.
To revise the list of rejected varieties.
To recommend such varieties as are worthy of general cultivation or are adapted to particular localities.
To add to the rejected list such as are unworthy of cultivation.
To appoint a temporary committee of seven on synonyms, who shall sit during the convention.

At their meeting in Philadelphia, Mr. J.J. Thomas from New York proposed that the name of the society be changed to the American Pomological Society. From their meeting in Philadelphia, they proceeded to meet every two years, beginning with their next meeting in Boston in 1854 and continuing until 1864 when they met in Rochester. They did not meet in 1866 but did meet in 1867 in St. Louis to resume the biennial schedule of meetings. Beginning in 1920, the Society began holding annual meetings, a tradition still being followed. Between 1852 and 1956, the transactions and proceedings of the Society were recorded as The Proceedings of the American Pomological Society. These bound volumes contain a rich detailed history of the Society. The annual presidential address, the financial report, committee reports, resolutions and sometimes, the Fruit Catalogue, all appeared in these bound proceedings.

Their initial meetings were largely devoted to the evaluation of fruit specimens brought to the meeting for comparisons and the discussion of the merits of the fruit varieties listed in their catalogue. In particular, the evaluation of pear cultivars was a lengthy discussion topic. They adopted rules at their convention that proved to be quite interesting, such as, "During the hours of the day the examination of the fruits shall engage the attention of the convention…. One specific variety only shall be examined at a time…."

When specimens of the fruit were present, the fruit were cut and evaluated by the members. This lead to a lively exchange, as each person would comment upon the fruit under examination. When more than a few people attended the convention, the noise could be quite deafening. Therefore, only those pomologists who had "loud clear voices made themselves heard" by the official reporter at the meeting. Some might say that the loudness of voice may have been a screening process to be a noted pomologist. As the Society matured, reason prevailed, and only a limited number of people were chosen to comment on fruit specimens.

One constant source of irritation to the members of the Society in the early years was the practice of unscrupulous nurseries taking an old cultivar, renaming it, and then boasting of their having exclusive rights to propagate and sell that cultivar. The members were constantly evaluating fruit cultivars so that they were not renamed and sold as new, improved cultivars. One instance cited in the proceedings was the case of a grape cultivar with three names, 'Jumbo,' 'Columbian Imperial,' and 'Columbian.' To make matters worse, this cultivar was no longer recommended to be planted.

The "reading" of papers on topics of concern to pomologists and fruit growers gradually came to take on a greater importance at the meetings. Presenters of these papers read like a who's who in early pomology: T.V. Munson of Texas, S.A. Beach from the New York Experiment Station in Geneva, W.T. Macoun from Canada, J.H. Gourley from Ohio, Paul C. Stark from Missouri, and Liberty Hyde Bailey of New York.

In accordance with one of the primary stated functions of the Society, fruit lists containing the names of desirable and undesirable varieties were published in The Catalogue of Fruits.

In later years, part of the catalogue's comprehensive lists of new fruits and nuts were prepared by the Committee on New Fruits and Nuts.

In 1944, Professor E.R. Lake published a complete alphabetical listing of new cultivar names appearing in the Society's annual reports from 1920 through 1943. This special publication, along with the New York Experiment Station series of Fruits of New York, contained all the cultivars of fruits ever introduced in this country to that date.

The catalogue later evolved into the Register of New Fruit and Nut Varieties as edited by R.M. Brooks and H.P. Olmo. Through its Catalogue of Fruits, the Society recognized that cultivars were the cornerstones of the fruit industry, and that many cultivars have specific climatic and soil requirements. The catalogue provided a record of recommended cultivars and brief descriptions. Through the catalogue, the work of the American Pomological Society did much to raise the standards by which fruit were judged and to discourage the cultivation of inferior cultivars. In the first 29 years of publication, the catalogue rejected 600 cultivars as being inferior, and by 1907 had rejected 1,626 cultivars.

The Society formulated the Code of Nomenclature as a guide for naming new cultivars. The code was used throughout the United States and Canada. The goal was to encourage developers of new cultivars to use names that were both appropriate and simple. Multiple names such as Hubbardston's None Such and Peach Pond Sweet apples, Connett Southern Early and Cook's Late White peaches, or Mrs. Pince's Black Muscat grape were discouraged from use in naming new cultivars.

In the 1920s, with the encouragement of the Society, Dr. J.K. Shaw of the University of Massachusetts, developed a descriptive key for the identification of apple cultivars. The keys were periodically updated, with the last being published in 1943. The key allowed the inspection of nursery beds to insure trueness to name.

The Society would frequently meet in conjunction with other horticulture-related groups. On the anniversary of their first 50 years, the Society met in Philadelphia at the request of the Pennsylvania Horticultural Society. The centennial anniversary meeting was held in St. Louis in partnership with the Missouri and Illinois State Horticultural Societies, the National Peach Council, the National Apple Institute, the American Society for Horticultural Science, and the American Association of Nurserymen. The sesquicentennial anniversary meeting was held in Charlotte, North Carolina, with the American Society for Horticultural Science.

The Society is the mother of a number of horticultural societies. The National Peach Council, the United States Apple Association, and the American Society for Horticultural Science all owe their beginnings to the members of the American Pomological Society and its sponsorship.

No history of the American Pomological Society is complete without mentioning Marshall Pinckney Wilder (1798-1886). He was elected president at the 1852 convention and held that office until 1886. Marshall Wilder was a strong president who helped nurture the Society in its infancy. Wilder was a distinguished amateur pomologist and floriculturist. He inherited a love of country life and chose farm work in preference to a college education. He became a prosperous Boston merchant and bought a home in Dorchester, Massachusetts, just outside of Boston. His avocation was in the evaluation of pear cultivars. He developed a pear orchard containing at one time 800 cultivars. During his life, he tested 1,200 cultivars of pears and exhibited 404 in 1873. He produced several new cultivars and introduced the 'Beurre d'Anjou' cultivar. He imported many fruits and flowers into this country. Some people may think the reason pears were discussed at great lengths during his presidency of the Society was due to his influence. In reality, it was just that pears were much more prominent in American horticulture at that time.

His greatest services to horticulture were connected with the Massachusetts Horticultural Society and the American Pomological Society. In recognition for his long service to the American Pomological Society, the "Wilder Medal" was established, which is given to pomologists who have contributed most to evaluation and improvement of cultivars of various kinds of fruit in this country. Upon his death in 1887, he bequeathed to the Society $1,000 for the support of future Wilder Medal Awards and a sum of $4,000 for the general use of the Society. This legacy necessitated that the Society be incorporated as a public organization in the state of Massachusetts in 1887. Wilder played not only an important role in Society affairs but was also active on a national level. He helped draft the Act legislation which created the land grant college system.

Today, the American Pomological Society continues to publish the *Fruit Varieties Journal* in four issues per year. Membership numbers approximately 900 fruit growers, industry members, research scientists, and extension personnel. The business office has been located at Penn State University at 103 Tyson Building, University Park, Pennsylvania, for more than 30 years. The Society continues to award Wilder Medals to individuals for their contributions to pomology. It also promotes interest in pomology in the next generation by sponsoring the annual U.P. Hedrick writing competition for college students.

The 'Delicious' Apple

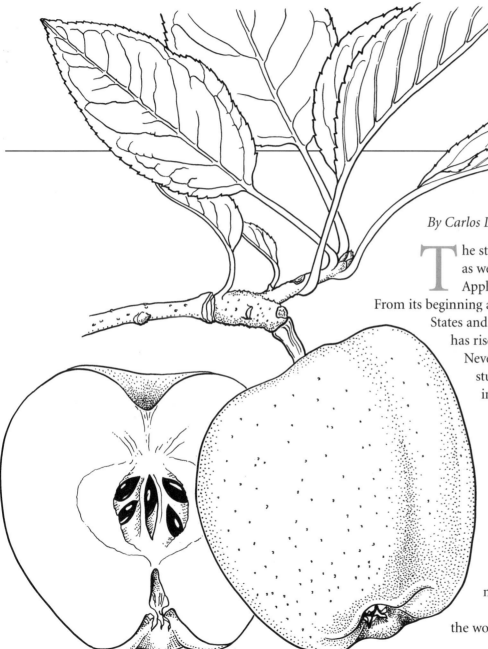

By Carlos D. Fear & Paul A. Domoto

The story of the 'Delicious' apple has been told many times and is almost as well known among apple enthusiasts as the legend of Johnny Appleseed. In Iowa, it represents an important part of our state heritage. From its beginning as a persistent unwanted seedling to its prominence in the United States and world fruit industry, the 'Delicious' apple, in particular its strains, has risen to become the most widely planted cultivar in the world (1). Never before in the history of pomology has one cultivar been so widely studied. It is appropriate to review the history and contribution of this important cultivar in the *Fruit Varieties Journal*.

It was in 1872 that Jesse Hiatt of Peru, Iowa, discovered a sprout arising from the roots of a 'Bellflower' seedling which previously had been twice cut down because it was growing out of the row. It is said that Hiatt remarked about the sprout which became the original 'Delicious' tree, "If thee must live, thee may." Hiatt realized when the sprout developed into a tree and began to fruit that it was not like the 'Bellflower' parent from which it came. The aroma was different, and the fruit had a distinctive shape with prominent calyx lobes. Hiatt was so impressed with the characteristics of the seedling that he named it 'Hawkeye' in honor of Iowa's nickname.

Jesse Hiatt so believed that he had discovered the best apple in the world that he tried for 11 years to convince others of its value by giving

FIGURE 1.
The original 'Delicious' apple tree sprouted again after dying back to the ground in the 1940 freeze. It is today surrounded by cornfields (1986 photo).

samples to friends and exhibiting it at fairs and fruit shows. It was not until 1893 that Hiatt's discovery found an admirer, Clearance M. Stark, at a fruit show at Stark Brothers Nursery in Louisiana, Missouri.

Mr. Stark had been reserving the name 'Delicious' for a variety worth the significance of the name, but not until tasting 'Hawkeye' had he found such a variety. The discovery and renaming of 'Hawkeye' almost failed to occur because the name and address of its exhibitor was lost at the show. Mr. Stark could only wait and hope that the fruit deserving the name 'Delicious' would be exhibited again the following year. Fortunately, Hiatt exhibited 'Hawkeye' again in 1894, and Stark immediately recognized it. Permission to propagate 'Hawkeye' was given to Stark Brothers Nursery, and henceforth, the variety was known as 'Delicious.'

Development of 'Delicious' as a variety

Several factors have contributed to the rapid development of 'Delicious' as a leading apple variety. Stark Brothers Nursery accelerated the development of 'Delicious' as a well-known variety by including trees of it as a free gift with nursery orders. The enthusiasm C.M. Stark had for the cultivar and the response of customers who had received free trees no doubt had a significant effect on stimulating extensive planting. By 1922, the annual value of the 'Delicious' crop was estimated at 12 million dollars.

'Delicious' as a cultivar possesses several characteristics which led to its rapid acceptance and planting. First, the apple was an extremely attractive cultivar, and its distinctive shape and prominent calyx lobes allowed it to become easily recognized by consumers. For the grower, 'Delicious' possesses good storage and shipping qualities, as well as being moderately easy to grow to high quality. Problems with fireblight, poor fruit finish, and blister canker known to other varieties were unknown to 'Delicious.' Selection of high-coloring strains has further enhanced its appearance and has allowed growers to harvest highly colored fruit before it becomes too mature.

As a tree, 'Delicious' and its strains have been accepted by growers because of its hardiness, good vigor, annual bearing tendency, precociousness (although somewhat low by present standards), and its ability to grow satisfactorily over a wide geographic area. Its popularity among consumers has overshadowed its regional problems of bitter pit, internal bark necrosis, dead spur disorder, additional tree training requirements, poor fruit set, variability in typiness, and union necrosis when propagated on certain rootstocks.

Strains of 'Delicious'

Strains or mutants tracing to the original 'Delicious' number over 100. From the beginning, when mutants were first discovered in 1919, strains with better coloring, different color patterns, and spur-type growth have been selected. Most of these have been either whole tree mutants or bud mutations discovered in commercial orchards.

'Richard' was one of the first strains found. It was found as a bud mutation of 'Delicious' in 1919 in an orchard in Monitor, Washington. 'Richard' differed from the parent 'Delicious' by having better fruit color and more blushed red color pattern.

'Starking' was found in 1921 in an orchard in Monroeville, New Jersey. It developed color much earlier than 'Delicious' and was a very significant mutation because it overcame one of the main faults of 'Delicious', that being a lack of good red color development until late in the season, sometimes when the fruit was too mature for good storage. 'Starking' became widely planted, and itself became the parent of more than 25 named strains.

The first spur type from 'Delicious' was 'Okanoma,' found in 1921 in Omak, Washington. 'Okanoma' did not find wide acceptance because of its poor coloring characteristics. However, many other spur types followed and are widely planted now. Fisher and Ketchie (1) discussed strains of spur and standard types. Spur types are distinctive in several ways from standard types by having less branching, flower initiation on short spurs instead of side shoots, and their tendency to grow more upright. Spur strains also bear earlier and generally have better skin color.

Although strain mutations of 'Delicious' have been selected for higher color, coloring pattern, and spur-type growth, other morphological, chemical, and physiological characteristics of the tree and fruit have occurred among the strains, and are well documented in the review by Fisher and Ketchie (1). Included among these characteristics are differences in fruit set, productivity, and fruit typiness. Due to these differences as well as color

"If thee must live, thee may."

Jesse Hiatt, of the original 'Delicious' tree.

and growth habit, studies comparing 'Delicious' strains are being conducted at several sites in the U.S. and elsewhere to determine which strains are best adapted to the various areas. Currently, over 40 strains are for sale in the nursery trade.

'Delicious' as a parent in breeding

'Delicious' and its strains have been used as parents by apple breeders in many countries. At least 33 named cultivars have 'Delicious' or one of its strains as a parent in their background. Some of these varieties involving 'Delicious' directly as a parent (e.g., 'Empire', 'Fuji'), or as a parent further back in their ancestry (e.g., 'Gala') have achieved worldwide recognition for their fine quality. Several others involving 'Delicious' as a parent are well-known and appreciated in certain regions of the country (i.e., 'Chieftain' and 'Jonadel' in Iowa, 'Melrose' in Ohio, and 'Regent' in Minnesota).

The 'Delicious' tree today

At the time Jesse Hiatt died in 1898, he and his variety were still unknown to the world. By 1922, however, Jesse Hiatt and 'Delicious' were well-known. In August of that year, a 50th year anniversary celebration of the discovery was planned, and a six-ton inscribed boulder-monument was dedicated and placed in the Winterset City Park, Winterset, Iowa. The original tree was still living, but difficulties in purchasing land right-of-way prevented dedicating the monument at the site of the tree.

The original 'Delicious' tree lived until the Armistice Day freeze in 1940, when temperatures dropped 50°F to near zero in 24 hours after a mild fall. With its incredible vigor, the tree sprouted again after being killed back to the ground and developed two sprouts which eventually fruited like the original tree. One of these trees still stands, proudly surrounded by cornfields, as the source of one of the greatest apple varieties ever discovered. The world should be thankful that Jesse said of the 'Delicious' seedling, "If thee must live, thee may."

Literature cited
1) Fisher, D.V. and D.O. Ketchie. 1981. Survey of literature on red strains of 'Delicious.' Wash. State Univ. College of Agr. Res. Cntr. Bulletin 0898.

Additional reading
Ferree, D.C., C.A. Morrison, and L.C. Shew. 1975. Red Delicious apples—which strain is best? Rpt. Ohio Agr. Res., Dept. Agr. Home Econ. Nat. Res. 60(2):19-22.

Lewis, C.I. 1922. The Delicious apple: its place in American pomology. Proc. Iowa State Hort. Soc. 57:87-90.

Maas, V. 1970. Delicious. *In* North American apples: varieties, rootstocks, outlook. Michigan State Univ. Press, East Lansing, pp. 45-68.

Miller, H.C. 1972. The Delicious apple, 100 years—1872-1972. Madison Co. Historical Soc., Winterset, Iowa.

Pellett, K. 1940. Jesse Hiatt and Delicious. Proc. Iowa. State Hort. Soc. 75:36-42.

Stark, P. 1922. The romance of the distribution of the Delicious apple. Proc. Iowa State Hort. Soc. 57:90-94.

Zeller, E.R. 1922. The Hiatts. Proc. Iowa State Hort. Soc. 57:85-87.

Authors Fear and Domoto were with the Department of Horticulture, Iowa State University, Ames, Iowa, when this article first appeared, January 1986.

The 'Bartlett' Pear

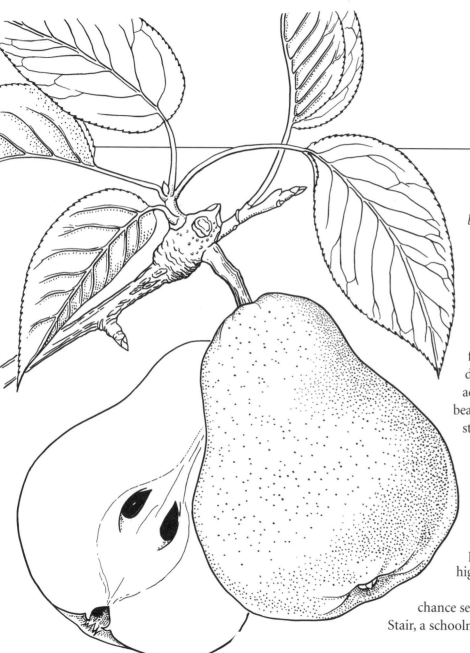

by Craig K. Chandler

'Bartlett' is the most important pear cultivar in the world. It accounts for 75 and 85% of the pear trees growing in the western and eastern United States, respectively (1,9). And, in Italy, the country with the highest production of pears, 'Bartlett' is second only to 'Passe Crassane' in percent of total output (2).

Hedrick (3) said of 'Bartlett,' "As with the leading variety of any fruit, the…meritorious character of this one is its great adaptability to different climates, soils, and conditions." In addition to its wide range of adaptation, 'Bartlett' has several other outstanding characteristics: It starts to bear relatively early, is productive, and can be grown successfully in either a standard or high-density orchard system (7,9).

While there are other cultivars that are considered superior to 'Bartlett' for eating fresh, 'Bartlett' is generally considered the standard of excellence for canning. The pear canning industry in the United States is based almost exclusively on 'Bartlett,' with nearly two-thirds of the 'Bartlett' crop being used for this purpose (9).

'Bartlett' is one of the few pear cultivars whose fruit contains a particularly high concentration of ethyl and methyl decadienoate esters— high-impact flavor compounds responsible for 'Bartlett's' unique flavor (6). Like the world's most important apple cultivars, 'Bartlett' originated as a chance seedling. According to Hedrick (3), this seedling was found ". . . by a Mr. Stair, a schoolmaster at Aldermaston, Berkshire, England. From him, it was acquired by

> *The flesh of 'Bartlett' pear is creamy white, melting, and juicy.*

a Mr. Williams, a nurseryman at Turnham Green, Middlesex, and as it was propagated and distributed by him, it became known by his name, although it is still known as Stair's pear at Aldermaston. It was brought to the United States in 1797 or 1799 by James Carter of Boston for Thomas Brewer, who planted the variety in his grounds at Roxbury, Massachusetts, under the name of Williams' Bon Chretien, by which name it was then and still is known both in England and France. In 1817, Enoch Bartlett, Dorchester, Massachusetts, became possessed of the Brewer estate, and not knowing its true name, allowed the pear to go out under his own. Henceforth, it was known in America as Bartlett."

'Bartlett' is commonly referred to as a summer pear because it is harvested from early July in the valleys of central California to late August and September in more northerly areas. The ripe fruit of 'Bartlett' usually has a pyriform shape and yellow skin with an occasional crimson blush. Its flesh is creamy white, melting, and juicy. The fruit has a storage life of 70 to 85 days in common cold storage.

Although 'Bartlett' has many outstanding traits, it is not without faults. Its biggest weakness is probably its susceptibility to fireblight caused by the bacterium *Erwinia amylovora* (Burr.) Winsl. et al. Another major weakness of 'Bartlett' is the tendency of its fruit to soften and ripen prematurely when exposed to cool night temperatures during the month before harvest (5).

'Bartlett' has been used extensively as a parent in pear breeding programs, and its named offspring include 'Gorham,' 'Aurora,' 'Highland,' 'Harvest Queen,' and 'Harrow Delight.' Many of 'Bartlett's' offspring resemble it in appearance, size, shape, and productivity, but very few produce fruit with the characteristic 'Bartlett' flavor (4). Various fruit color and russet mutants of 'Bartlett' have been recognized and propagated. Red-fruited mutants of 'Bartlett' include 'Cardinal Red,' 'Sensation,' and 'Rosired.' Despite the fact that 'Bartlett' was discovered almost 200 years ago, none of its offspring or sports seem destined to replace it as the world's preeminent pear cultivar.

Literature cited

1) Chase, Patrick C. 1985. Review of pear production in the East. *Compact Fruit Tree* 18:127-30.

2) Fedeghelli, Carlo. 1985. Prospect of apple and pear industry in Italy. *Compact Fruit Tree*, Suppl. to Vol. 18.

3) Hedrick, U.P. 1921. The pears of New York. New York Agr. Exp. Sta.

4) Lamb, R.C. 1982. Pear breeding in New York State, *In:* T. van der Zwet and N.F. Childers (eds.), *The Pear*. Hort. Publ. Gainesville, FL. pp. 171-79.

5) Lombard, P., J. Hull, Jr., and M.N. Westwood. 1980. Pear cultivars of North America. *Fruit Var J.* 34(4):74-83.

6) Quamme, H.A. and P.B. Marriage. 1977. Relationships of aroma compounds to canned fruit flavor among several pear cultivars. *Acta Hort.* 69:301-306.

7) van den Ende, Bas and David Chalmers. 1983. The trellis as a high density system for pear trees. *HortScience* 18(6):946-47.

8) Westwood, Melvin N. 1978. *Temperate-zone pomology*. W.H. Freeman and Company, San Francisco.

9) Williams, Max W., H. Melvin Covery, Harold Moffitt, and Duane L. Coyier. 1978. Pear production. USDA Agriculture Handbook No. 526.

Author Chandler was with the Department of Horticulture, Ohio State University, Wooster, Ohio, when this article originally appeared, April 1986.

The 'Valencia' Orange

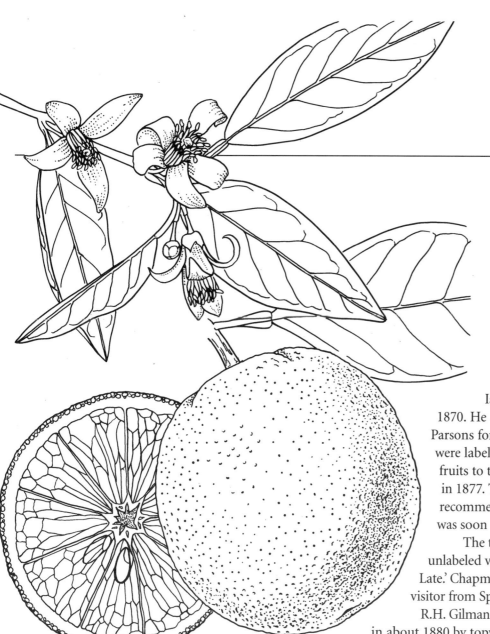

by R.K. Soost

In spite of its name, the 'Valencia' orange [*Citrus sinensis* (L.) Osbeck] does not appear to have been introduced into the United States from Valencia, or even Spain. Although the evidence isn't conclusive, it appears to trace to Portugal. The Portuguese variety 'Don Joao' appears to be similar or identical to 'Valencia' (4).

The English nurseryman Thomas Rivers imported a cultivar from the Azore Islands and listed it in 1865 under the name 'Excelsior' (2). He sent trees from England to S.B. Parsons of Long Island, New York, and to General Sanford of Palatka, Florida, in about 1870. He also sent trees to S.B. Chapman of San Gabriel, California, in 1876. Parsons forwarded trees to E.H. Hart of Federal Point, Florida. Sanford's trees were labeled "Brown," but Hart's trees arrived without labels. Hart submitted fruits to the nomenclature committee of the Florida State Horticultural Society in 1877. The committee named it 'Hart's Tardiff,' published a description and recommended its trial. The name was later changed to 'Hart's Late' or 'Hart.' It was soon discovered that 'Sanford's Brown' and the 'Hart' were identical (6).

The trees received in California by Chapman were in a shipment of several unlabeled varieties, one of which proved to be late maturing and was called 'Rivers Late.' Chapman later changed the name to 'Valencia Late' at the suggestion of a visitor from Spain, who thought it similar to the 'Naranja Tarde de Valencia' of Spain. R.H. Gilman established the first commercial orchard in California near Placentia in about 1880 by topworking an existing five-acre planting. However, the first car lot of

The origin of 'Valencia' is unknown.

Author R.K. Soost was with the Department of Botany and Plant Sciences, University of California, Riverside, when this article originally appeared, July 1986.

fruit was shipped in 1880 by Colonel J.R. Dobbins from a planting that had been topworked with buds, supposedly 'Washington' navel, purchased from S.B. Chapman. However, the buds had been mistakenly cut from Chapman's 'Rivers Late' trees. Through this fortunate error, which for a time threatened to end in a lawsuit, Colonel Dobbins became the first extensive shipper of 'Valencia' oranges in California (1). It was several decades before it was determined that 'Hart's Tardiff' and 'Valencia Late' were identical.

Although the introduction of the 'Valencia' into the United States clearly is traceable to Thomas Rivers and in turn to the Azore Islands, the origin of the 'Valencia' is unknown. Lee and Scott (5) suggested that some oranges grown in southwest China were very similar to the 'Valencia' and may have been taken to Europe by early Portuguese or Spanish voyagers. Cooper (3) more recently suggested that the very old Portuguese cultivar 'Selata' is identical to 'Valencia' and 'Don Joao,' except for earlier maturity. The 'Selata,' according to Cooper, also resembles the 'Sukka of Shanton' from China. He suggests that the 'Selata' originated from 'Sukka' and the 'Valencia' in turn from 'Selata.' However, concrete evidence is lacking.

Whatever its origin, the 'Valencia' orange has become the most widely grown cultivar of the sweet orange group because of its adaptation to a wide range of climatic conditions. It is of major importance in California, Florida, South Africa, Australia, and in the Mediterranean area. It is also grown in Brazil and other South American countries.

Various budlines have been selected and established in the various countries where 'Valencia' is grown. Many of these have been named, although most may not differ from the original. Slight differences in tree growth and fruit characters have been detected among some of the selections.

In addition to bud selections of 'Valencia', several other cultivars throughout the world appear to be identical to Valencia. These include 'Lue Gim Gong' and 'Pope' from Florida, 'Natal' of Brazil, 'Calderon' of Argentina, and 'Harward' of New Zealand (4). The 'Lue Gim Gong' and the 'Harvard' originated as seedlings of 'Valencia' and are undoubtedly of nucellar origin.

Literature cited

1) Clark B.O. and W.A. Johnstone. 1937. Introduction of the 'Valencia' orange into California. California Citrograph 22:509, 531-533.

2) Coit, J.E. 1915. *Citrus fruits.* McMillan Co. New York.

3) Cooper, W.C. 1982. *In search of the golden apple.* Vantage Press, New York.

4) Hodgson, R. W. 1967. Horticultural varieties of citrus. In: Reuther, W., H.J. Webber, and L.D. Batchelor (eds.) *The Citrus Industry,* Revised Ed., Vol. 1., Univ. Calif. Div. Agri. Sci., Berkeley. pp. 431-591.

5) Lee, H.A. and L.B. Scott. 1920. Are 'Valencia' oranges from China? *J. Heredity* 11:329-333.

6) Webber, H.J. 1948. Cultivated-varieties of citrus. In: Webber H J. and L.D. Batchelor (eds.) *The Citrus Industry,* Revised Ed., Vol. 1., Univ. Calif. Press, Berkeley and Los Angeles. pp. 475-668.

The 'Concord' Grape

by G.A. Cahoon

Domesticated grapes having their origins in North America belong to a relatively few species, the most predominant of which is *Vitis labrusca*, Linnaeus, the North American Fox grape. 'Concord' is perhaps the most famous of this group and certainly the most widely grown. Its wide adaptability has resulted in its production in almost every grape-growing state. True to its parentage, 'Concord' has an epidermis that easily separates from the flesh or pulp and seeds that do not. This "slipskin" characteristic is in contrast to European or *Vitis vinifera* grapes whose epidermis adheres to the flesh of the fruit. Although 'Concord' is commonly accepted and described as pure *V. labrusca*, there is a reasonable probability that it is a hybrid with another species. In fact, most of the older American-type grapes are thought to involve more than one species. Therefore, *Vitis labruscana*, L.H. Bailey, is now commonly used to designate American grape cultivars having labrusca parentage (2, 3). Botanical characters of 'Concord' include continuous tendrils and a white-to-rusty tomentum on the under side of the leaf blade (2, 7). Other well-known cultivars which fall into this group include 'Catawba,' 'Niagara,' 'Isabella,' and 'Ives.' Examples of North American grapes in which other species characteristics predominate are: 'Norton' (*V. aestivalis* x *V. labrusca*), 'Delaware' *(V. aestivalis* x *V. bourquina*—*V. labrusca* x *V. vinifera)*, and 'Elvira,' 'Clinton' *(V. vulpina* x *V. labrusca)* (7).

The origin of 'Concord' is well documented and has been told many times. Certainly, it has a history that compares to any of our famous and outstanding fruit and must be retold here in order to properly complete our story.

In the early history of our country, many attempts were made to introduce the European or *V. vinifera* cultivars into the eastern United States. For well-documented reasons (winter temperatures, phylloxera, and disease), all were doomed to failure. Winemakers struggled with the abundant—but poor quality—wild, native grapes. The rapid spread of 'Catawba' throughout the East and Midwest testifies to the need for a better quality wine grape (4, 5, 8).

Thus, it would not be unusual to find that Ephraim Bull, a grape connoisseur from Concord, Massachusetts, had 'Catawba' growing in his garden. As the story goes, he dug up a wild, native labrusca grape growing nearby and planted it in his garden where other grapes were growing, including 'Catawba.'

In the fall of 1843, he gathered seed from the fruit of this wild vine and planted it. Among the seedlings that grew was one of outstanding vigor and character. Initial production was obtained in 1849.

Eventually it was named 'Concord.' As quoted by Mr. Bull himself in a letter to C.M. Hovey, editor of the *Magazine of Horticulture*, in January 1854: "The Concord grape is a seedling, in the second generation, of our native grape… being at that time the only seedling I had raised which showed a decided improvement on the wild type…. The seedling from which the 'Concord' was raised grew near a 'Catawba,' and it is quite possible was impregnated by it, it having the flavor of that variety" (11).

As mentioned previously, 'Concord's' characteristics are strongly labrusca. However, it has upright stamens (perfect flowers) and is self-fertile, which is not typical of some wild labruscas. The character of many of the seedlings also leaves some doubt.

History indicates that 'Concord' was exhibited for the first time on the 20th of September, 1853, before the Massachusetts Horticultural Society on Boston Common—three years after it had produced its first fruit. In 1854, it was introduced by Hovey and Company of Boston. During this period, horticultural societies were in frequent contact with one another on new fruit varieties and cultural practices. Also, the numerous farm journals of this period provided an additional source of information.

From this time, 'Concord' spread quite rapidly throughout most of the eastern and midwestern states. By 1867, the Ohio Horticultural Society was writing about the extensive plantings of the 'noble Concord' in Ohio and Missouri (6). By the mid 1870s, more 'Concords' were planted in the Northeast than all other varieties put together (5). It had become the outstanding grape for both fresh and processed use. Fruit was shipped from the grape belts of the Lake Erie region to most of the major U.S. cities. Yet, in the same breath, it was berated as being no good for shipping, wine, etc.

In Ohio, Prohibition had a rather interesting effect on grape acreage and grape varieties. 'Concord,' the predominant grape for table use, but thought of as the "other grape" for winemaking, now came into prominence. 'Catawba,' although a superior wine grape, was not as well-known to the home winemaker. Thus, a transition occurred. 'Concord' acreage increased, while 'Catawba' vineyards were being abandoned or quickly replanted to 'Concord.' 'Concord' grape juice was in demand and sold with specific instructions on how to avoid fermentation, which was certainly the reason for making the statement in the first place.

'Concord' has been used as a parent in many breeding programs. As stated by Hedrick (7), 'Concord' is the progenitor of many other varieties. Among the more familiar names are 'Moore's Early,' 'Diamond,' 'Worden,' 'Niagara,' 'Brighton,' 'Martha,' 'Cottage,' 'Highland,' and many others.

Today, 143 years after its creation, 'Concord' still stands as the predominant grape of the East. According to a review of grape production in the United States, more than 70% of the grapes produced in the Northeast, North Central states, and Northwest are of this cultivar (1, 4, 9, 10). The flavor of 'Concord' remains as the standard of quality for all processed grape products from jams and jellies to soft drinks.

Literature cited

1) Ahmedullah, M. 1980. Grape Growing and Cultivar Review of the Pacific Northwest. *Fruit Var. Jour.* 34(3):61-65.

2) Bailey, L.H. 1922. *The Standard Cyclopedia of Horticulture.* The Macmillan Co., New York. p. 3490.

3) Bailey, Liberty Hyde, and Ethel Zoe Bailey. *Hortus Third, A Concise Dictionary of Plants Cultivated in the United States and Canada.* Macmillan Publishing Co., Inc., New York. p. 1163.

4) Cahoon, G.A. 1980. Grape Production in Four North Central States and Kentucky. *Fruit Var. Jour.* 34(3):54-59.

5) Cahoon, G.A. 1984. The Ohio Wine Industry from 1860 to the Present. *Jour. Amer. Wine Soc.* 16:82-86, 94.

6) Warder, G.A. 1867. First Annual Report of the Ohio State Hort. Soc. pp. 7, 15.

7) Hedrick, U.P. 1908. The Grapes of New York. Fifteenth Annual Rpt. State of New York. Vol.3 Part 11, pp. 219-222.

8) McGrew, John R. 1984. A Brief History of Grapes and Wine in Ohio to 1865. *Jour. Amer. Wine Soc.* 16:38-41.

9) Moore, J.N. 1980. Cultivar Situation in Arkansas, Missouri, Oklahoma and Texas. *Fruit Var. Jour.* 34(3):59-61.

10) Pool, Robert M. 1980. Northeast Region. *Fruit Var. Jour.* 34(3):50-54.

11) Tukey, H.B. 1966. The Story of the Concord Grape. *Fruit Varieties and Horticultural Digest.* 20(3):54-55.

> *By the mid 1870s, more 'Concords' were planted in the Northeast than all other varieties put together.*

Author Cahoon was with the Department of Horticulture, Ohio State University, when this article originally appeared, October 1986.

The 'Searles' Cranberry

by Elden J. Stang

The commercial cranberry, *Vaccinium macrocarpon* Ait., is native to bogs of the northeastern United States and southern Canada. 'Searles' is among the more than 100 cranberry selections from the wild that were propagated and eventually introduced as cultivars over more than 150 years of commercial cultivation of this species.

In addition to many of these named selections, seven cranberry cultivars from a USDA breeding program and one cultivar from Washington State University were recently compiled by Dana (3). The descriptions compile information from partial listings by Hedrick (4), Chandler and Demoranville (2), and Brooks and Olmo (1). Additional cultivars were added for which information was obtained from records of the Wisconsin Cranberry Experiment Station for 1903-1917.

Hedrick, in 1922, referring to 'Searles Jumbo,' described the fruit as being of the Jumbo type, an apparent reference to various native cranberry selections noted for large fruit size. The 'Searles Jumbo' berry is olive-shaped, uniform, bright crimson red when ripe and fully exposed to light at the top of the vine canopy. Fruit is uniformly large, not glossy, only rarely mottled. The cultivar is designated as "medium" or mid-season ripening, harvested in late September or early October, after 'Ben Lear,' in Wisconsin.

'Searles Jumbo' was discovered and propagated by Andrew Searles (1852-1933) in his native bog near Walker, west of the city of Wisconsin Rapids, at that time named Grand Rapids, Wisconsin. Neither the exact date nor the circumstances of the discovery are adequately documented, although Peltier indicates vines were collected by Searles in 1893 and placed in nursery plots at the nearby Cranberry Experiment Station in 1894 (5). Larger plots remained at the experiment station until it closed in 1917 due to lack of funds. Of the named cultivars originating in Wisconsin, 'Searles Jumbo' was often noted as the outstanding selection in the test plots. Eventually, the plots were plowed up and replanted to commercial cultivars.

Reportedly, the cultivar was slow to be planted as a result of a slow increase in vines and the high cost of cuttings. By 1928, 'Searles Jumbo' acreage in Wisconsin had increased to 371 acres and doubled by 1939 to become the leading cultivar in Wisconsin. By 1956, 2,250 acres of 'Searles Jumbo' (58% of the total 3,900 acres of cranberries in Wisconsin) were in production (8). Prior to 1930, planting stock of 'Searles Jumbo' originated mostly from the Searles bog. Subsequently, rapid increases in acreage resulted from the greater quantity of vines for sale by other growers.

Some confusion exists as to the correct spelling for the cultivar name. In 1905, in his personal notes on cultivars (unpublished), O.G. Malde, superintendent of the Cranberry Experiment Station lists the name as 'Searls Jumbo.' According to Clarence A. Searles, grandson of the originator, spelling of Andrew's family name changed to Searles about 1918 (personal communication). A concerted effort was made by Andrew Searles to change the cultivar name to 'Searles Jumbo' in usage and publications after that time. Hence, Hedrick's reference in 1922 to 'Searles Jumbo' (4). The term 'Jumbo' is now rarely used; by usage, 'Searles' has become the accepted name in the trade.

Peltier, in 1970, summarizing his own and grower observations, characterized 'Searles' as a "rapid grower," prolific in blossoming, with large berries of a deep red color (5). Less favorable characteristics he noted included susceptibility to field and storage rots, especially end rot (*Godronia cassandrae* Peck), also known as stem rot. The cultivar was generally classified as having poor keeping quality due to its thin fruit skin and rather deep stem pit, easily wounded when pulled from the vine. Storage losses as high as 25% or more were noted in some seasons. At present, this susceptibility to storage losses is of little concern, as most Wisconsin berries are machine harvested, cleaned, sorted, and frozen in a period of one to three days for subsequent processing.

As with other species, cranberry cultivars selected in a specific area may be best adapted and most productive within a narrowly defined geographic region. Except for the limited acreage in Oregon and Canada, 'Searles' has become the predominant cultivar only in Wisconsin.

In 1968, the latest official survey of cranberry cultivars by acreage, made by the Wisconsin Statistical Reporting Service, noted 7,001 acres of cranberries in Wisconsin (7). Of that acreage, 61% was planted to 'Searles.' Despite some planting of other promising cultivars in the mid 1980s, unofficial estimates indicate 'Searles' currently remains the predominant cultivar, approximately 60% of the total of nearly 8,000 acres (6).

Cranberry production per unit area in Wisconsin more than doubled from 7,070 kg/ha in 1950 to 18,140 kg/ha in 1984. This dramatic increase in productivity resulted from rapid

> *As with other species, cranberry cultivars selected in a specific area may be best adapted and most productive within a narrowly defined geographic region.*

grower adoption of improved cultural management and improved fertilizer and pest control practices, including weed control. Throughout this period, 'Searles' remained predominant and consistently the most productive cultivar. In 1985, cranberries contributed more than $350 million to Wisconsin's total agribusiness economy. For more than 50 years, the 'Searles' cultivar indisputably has ranked among the greatest contributions to the growth of this important, unique agricultural industry in Wisconsin. Andrew Searles perhaps never fully envisioned the future impact of his fortuitous selection of a native cranberry in 1893. We are indebted to him for his ambition and foresight in selecting and early testing of this important, productive cranberry cultivar.

Literature cited

1) Brooks, R.P. and H.P. Olmo. 1972. Register of new fruit and nut varieties. 2nd edition, University of California Press, Berkeley.

2) Chandler, F.B. and I.E. Demoranville. 1958. Cranberry varieties of North America. Mass. Agr. Exp. Sta. Bull. 513, 23 pp.

3) Dana, M.N. 1983. Cranberry cultivar list. *Fruit Var. Jour.* 37(4):88-95.

4) Hedrick, U.P. 1922. Cyclopedia of hardy fruits. The Macmillan Co., New York.

5) Peltier, G.L. 1970. A history of the cranberry industry in Wisconsin. Harlo Press, Detroit.

6) Stang, E.J. and M.N. Dana. 1984. Wisconsin cranberry production. *HortScience* 19(4):478, 607.

7) Wisconsin Statistical Reporting Service. 1968. Cranberries in Wisconsin, a survey. USDA, Wisconsin Department of Agriculture, Trade and Consumer Protection, Madison, Wisconsin.

8) Wisconsin State Department of Agriculture. 1957. Cranberries of Wisconsin: Supplies, Varieties, Markets. Special Bulletin No. 70, 39 pp.

Author Stang was with the Department of Horticulture, University of Wisconsin, Madison, Wisconsin, when this article originally appeared, January 1987.

The 'Redhaven' Peach

by A.F. Iezzoni

'Redhaven,' the most extensively planted freestone peach cultivar in the world, was the most notable of many contributions to fruit growing made by Stanley Johnston, peach breeder and director of the Michigan State University Experiment Station at South Haven from 1920 until his death in 1969. Before 'Redhaven' appeared, it generally had been accepted that the "perfect" peach color was a golden undercolor with 25% red blush. 'Redhaven,' the first of the red-skinned peach cultivars, was 75 to 90% red, setting a new standard for peach color and commanding premium prices.

In addition, 'Redhaven' had an unprecedented ability to adapt in all the important freestone peach growing areas of the world. Three characteristics of 'Redhaven' include vigor, high productivity, and above-average cold hardiness. However, thorough and early thinning is required to obtain peaches of large size. Ripening 30 days before 'Elberta,' the fruit's brilliant red color develops while the flesh is still firm.

In the early 1900s, the 'Elberta' peach completely dominated the peach cultivar list in practically every peach-growing state in the nation. As a result, orderly marketing of peaches was practically impossible. All available cultivars maturing before 'Elberta' were too soft for commercial handling and shipping. Stanley Johnston began breeding peaches in 1924 when 85% of Michigan's peach crop was 'Elberta.' His primary goal was to develop early ripening, firm-fleshed cultivars.

Since the peach is native to China, the Michigan program and others like it in other states used parents derived from 'Chinese Cling.' 'Chinese Cling' resulted from a seed received in 1850 by Henry Lyons of Columbia, South Carolina. It probably originated from the southern group of Chinese cultivars which were

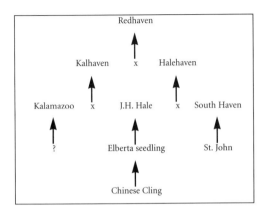

FIGURE 1.
The pedigree of 'Redhaven' peach.

Author Iezzoni was with the Department of Horticulture, Michigan State University, East Lansing, Michigan, when this article originally appeared, April 1987.

well adapted to warm, moist conditions in the southern United States (2). 'Elberta' originated as a seedling of 'Chinese Cling' on the S.H. Rumph farm in Georgia in 1840, and by 1916, it comprised approximately 80% of the freestone peach production in the United States. 'J.H. Hale,' which was first distributed in 1912, is a chance seedling, probably selfed, of 'Elberta,' found by J.H. Hale in Glastonbury, Connecticut.

The first crosses Johnston made in 1924 used the self-sterile 'J.H. Hale' as the female parent (Fig. 1). Two chance seedlings found in Michigan, 'South Haven' and 'Kalamazoo,' proved to have outstanding progeny when crossed with 'J.H. Hale.' The first selection, 'Halehaven,' was released in 1932, the second, 'Kalhaven,' in 1936. 'Halehaven' ripened 14 days and 'Kalhaven' 4 days before 'Elberta.' 'Halehaven' and 'Kalhaven,' while still numbered selections, were crossed in 1930, and the progeny 'Redhaven' was named and released 10 years later. According to Johnston (1), 'Redhaven' derived large fruit size, firm flesh, tough skin, bright yellow ground color, and a moderate amount of bright red skin color from 'J.H. Hale;' earliness, hardiness, and a bright red skin color from 'South Haven;' and, hardiness and a strong tree structure from 'Kalamazoo.'

Since its release in 1940, 'Redhaven' has been used extensively in breeding programs, and some of its progeny include 'Candor,' 'Madison,' and 'Pekin.' It is also in the pedigree of numerous cultivars, including 'Cresthaven,' 'Sweethaven,' 'Newhaven,' 'Jayhaven,' 'Clayton,' 'Correll,' 'Ellerbe,' 'Hamlet,' and 'Harrow Beauty.' 'Garnet Beauty' is an early ripening bud mutation of 'Redhaven.'

As early as 1960, 'Redhaven' accounted for large acreages in the United States, Canada, France, Italy, Yugoslavia, Turkey, other Mediterranean countries, some South American countries, and New Zealand. In a brief description of commonly grown peach cultivars written about 1960, Johnston simply wrote: "Redhaven matures about 30 days before Elberta. Little needs to be said about this variety. It is the leading variety of its season in the U.S. and is doing well in many other parts of the world." To commemorate Johnston's contributions, the state of Michigan erected an historical marker at the South Haven Station in 1966. On it are listed the eight yellow-fleshed freestone cultivars which Johnston bred in the course of 45 years of dedicated service to peach growing in America and the world.

Literature cited

1) Johnston, S. Breeding the Haven series of peach varieties. Proceedings XVIth International Hort. Congress, Brussels, Belgium, pp. 83-86.

2) Scorza, R., S.A. Mehlenbacher, and G.W. Lightner. 1985. Inbreeding and coancestry of freestone peach cultivars of the eastern United States and implications for peach germplasm improvement. *J. Amer. Soc. Hort. Sci.* 110:547-552.

The 'Napoleon' Sweet Cherry

by Susan K. Brown

Published reports of this white, sweet cherry cultivar are filled with praise befitting its royal name. It has been described as attractive, productive, of good size, excellent quality, and possessing admirable processing characteristics (1, 3). 'Napoleon' is of unknown origin. However, Hedrick cites reports of its culture as early as 1667, and notes its fine reputation in many European countries by the early eighteenth century. 'Napoleon' has many synonyms in other languages, but Hedrick believed that a Belgian by the name of Parmentier gave it the name of the famous emperor in 1820. The name 'Wellington' was substituted in England during the time Napoleon was not in favor, but the name was never widely used (3).

In local regions of America, the 'Napoleon' cherry is often called the Oxheart, but its most common synonym is 'Royal Ann.' This name was given to 'Napoleon' by Seth Lewelling, who had brought the cultivar across the country to Oregon, but lost the label along the way. The name 'Royal Ann' is commonplace on the West Coast.

'Napoleon' was placed on the American Pomological Society's Fruit List in 1862, after at least 40 years of performance in American orchards (3). It still represents a major cultivar because of the ease with which it can be bleached for maraschino cherry production, and the fine quality of its canned products.

Acreage of 'Napoleon' is decreasing in certain areas of this country, often in association with a reduction in that region's brining industry. However, it still represents 8% of the sweet cherry acreage in California (5), and 2.6% of all sweet cherry trees 11 years and older in Ontario, Canada (6), to cite just a few statistics. In Europe, it is a major cultivar

> *In local regions of America, the 'Napoleon' cherry is often called the Oxheart, but its most common synonym is 'Royal Ann.'*

in Czechoslovakia, Greece, and West Germany. It is also important in Japan, South Africa, and Australia. Ten European countries list 'Napoleon' in their most important cultivars for young orchards; this rating is second only to 'Van' (12 countries) and far exceeds the rating for 'Bing,' which only six countries list (2).

Although 'Napoleon' is a fine cultivar, certain attributes could be improved. Greater resistance to rain-induced fruit cracking, brown rot, bacterial canker, silver leaf infection, and western-X disease would be desirable. Fortunately, 'Napoleon' has been widely used as a parent in most sweet cherry cultivar improvement programs.

The New York State Agricultural Experiment Station released two selections, 'Gil Peck' and 'Sodus,' both of which resulted from the cross of 'Napoleon' by 'Giant.' Matthews stated that 'Napoleon,' 'Noble,' and 'Schrecken' were among the most successful parents in his breeding program, producing high percentages of useful offspring. 'Napoleon' was the male parent of both 'Merton Bigarreau' and 'Merton Late,' and the female parent of 'Merton Crane.' 'Merton Bigarreau' is now an important cultivar in England (4).

'Napoleon' is found in the pedigree records of many other cultivars, including the commercially important 'Lambert' (1). Unfortunately, 'Lambert' has the same susceptibility to rain-induced fruit cracking as 'Napoleon.' If this problem could be reduced, or eliminated in present or future 'Napoleon' offspring, then the prospect for a new "royal" leader in the cherry industry of the future would be very bright.

Literature cited

1) Christensen, J.V. 1977. A survey of the world production of cherries. Report State Research Station. Blangstedgaard, Denmark.

2) Coe, F.M. 1935. Cherries of Utah. Utah Agr. Exp. Sta. Bull. 253:41-43.

3) Hedrick, U.P. 1915. The Cherries of New York. Report of the New York Agricultural Experiment Station for 1914; Part II.

4) Matthews, P. 1973. Recent advances in breeding sweet cherries at the John Innes Institute. In: *Fruit present and future*. Volume 2. Royal Horticultural Society.

5) Nuckton C.F. and W.E. Johnston. 1985. California tree fruits, grapes and nuts: location of acreage, yields and production, 1946-1983. Giannini Foundation Information Series 85-1. University of California.

6) Tehrani, G. 1985. Cherry cultivars. In: *Fruit cultivars: A guide for commercial growers*. Publication 430. Ontario Ministry of Agriculture and Food.

Author Brown was with the Department of Horticultural Sciences, New York State Agricultural Experiment Station, Cornell University, Geneva, New York, when this article originally appeared, July 1987.

The 'Golden Delicious' Apple

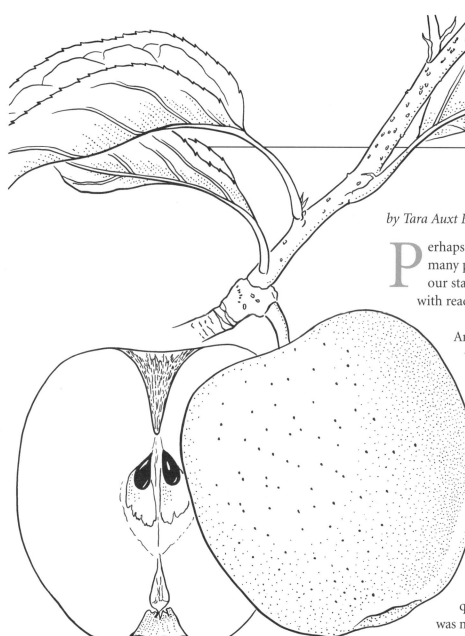

by Tara Auxt Baugher & Steve Blizzard

Perhaps no other West Virginian has touched the lives of so many people in so many places as the 'Golden Delicious' apple. With the pride typical of people from our state, we welcome this opportunity to share the story of the 'Golden Delicious' with readers of the *Fruit Varieties Journal*.

In the autumn of 1905, on a 36-acre farm in Clay County, West Virginia, Anderson Mullins noticed a precocious seedling with large yellow apples, unlike any variety seen before. Little did this humble hillside farmer realize that his discovery would one day change the course of world apple production. The golden apple tree located on a hill near Porter's Creek has since become the mother of millions of such trees throughout the globe.

For nine years after he first observed the tree, Anderson Mullins harvested an abundant crop annually, even when other trees in the family orchard were barren. The apple kept in good condition in the Mullins' cellar until April. Anderson called his discovery, which was at this time known only among neighbors and friends, 'Mullins Yellow Seedling.'

In the spring of 1914, Anderson Mullins sent three apples by parcel post to Stark Brothers Nursery, Louisiana, Missouri, modestly describing the observations he had made on the bearing habit of the tree and the quality of the fruit. When Lloyd Stark opened the small package of fruit, he was not very excited, because yellow apples had never been big sellers. Nevertheless,

FIGURE 1.
Senator Jennings Randolph and Paul Stark, Sr., joined other horticulture VIPs in planting a 'Golden Delicious' to commemorate the 50th anniversary of discovery. The list of distinguished horticulturists who attended included Paul Stark, Edwin Gould, L.P. Batjer, H.B. Tukey, Bill Luce, A.J. Heinicke, and delegates from Italy, France, Belgium, Great Britain, Mexico, Lebanon, Australia, New Zealand, South Africa, and Canada.

Lloyd and Paul Stark decided to sample a slice of one of the apples. The apple, which had been stored in Mullins' cellar all winter, was still firm, but what was even more impressive was the spicy flavor!

Quickly realizing the importance of the discovery, Paul Stark, an aggressive pomologist, set out on a journey that would later be called "The Trail of the Golden Apple." He told of his travels to West Virginia in a story often printed in his catalogs:

…he (Paul Stark) found he could reach Odessa only by a narrow-gauge railroad, and at the end of the line, he would have to travel by horseback. Stark rode on through the mountains and finally spotted a mailbox with the Mullins name on it. He knocked on the cabin door, but nobody answered. He was about to give up when he decided to have a look around. There was a small neglected orchard back of the house with gnarly apples, but high up on the side of the mountain back of the house, his eye suddenly caught a glint of golden yellow. He scrambled up the rocky slope, and there was the original tree loaded with big golden apples. He was biting into one when a mountaineer appeared from nowhere. Stark identified himself, saying, "That's some apple."

"Name's Mullins," the man said, "I sent you some." (2)

Stark found what he was looking for—an exciting new apple that would become a gold mine to the fruit industry. "It just growed there," Mullins told Stark. It was a naturally occurring chance seedling. Later research indicated that it grew from a 'Golden Reinette,' which was pollinated by a 'Grimes Golden,' another West Virginia discovery probably planted by Johnny Appleseed (John Chapman).

Stark Brothers paid $5,000 for the original 'Golden Delicious' tree. For thirty years, the Starks sent Bewel Mullins, Anderson's nephew, $100 a year to maintain the tree in good health as long as possible. Paul Stark corresponded with Bewel regularly to make sure that the tree was properly fertilized and sprayed. He also asked Bewel to keep accurate written and photographic records of the tree, paying close attention to any indications of a weakness.

When the Starks purchased the Mullins's tree, they also acquired a 900-square-foot tract of the hillside on which the tree stood. To protect their valuable real estate, the Starks built a woven-steel cage around the original 'Golden Delicious' tree. The door of the cage was secured by a lock, and a battery-powered burglar alarm was attached to a lead-enclosed wire which was strung from the cage to the Mullins's farmhouse. The alarm was activated only once, when the family cow strayed too close to the compound.

Research conducted at West Virginia University by H.E. Knowlton established the fact that the 'Golden Delicious' was equal to the old 'Ben Davis' and the 'Grimes' as a pollinizer for other apple varieties. The female element is equally as virile as the male side. Fruit is borne on one-year twigs and terminals in addition to fruit spurs, and record fruit loads are produced two to three years after planting. The tree is easy to manage and is adaptable to a wide range of soils and climatic conditions. These attributes make the 'Golden Delicious' a popular breeding parent. 'Mutsu' (Crispin), 'Jonagold,' and 'Gala' are some of the successful crosses. The original tree bore high quality, heavy crops for 50 years.

The primary weakness of the 'Golden Delicious' is that the skin tends to russet in humid climates. Another West Virginian, Henry W. Miller, Jr., offered a $5,000 reward for the discovery of a russet-free 'Golden,' but no claims were ever made. Evaluations by J.N. Cummins, P.L. Forsline, and R.D. Way showed that 'Smoothee' (Gibson) has the least russet of the currently available strains (1).

The new yellow cultivar had all the characteristics of a "winner." The American Pomological Society awarded the Wilder medal to the 'Golden Delicious' in 1921. Fifty years after the discovery of the 'Golden Delicious,' a ceremony was held in commemoration of the introduction of this variety to the rest of the world. United States senators, the governor of West Virginia, faculty of West Virginia University, ambassadors from foreign countries, and representatives of the United States Department of Agriculture paid homage to the tree that was discovered on Porter's Creek. Edwin Gould, superintendent of the West Virginia University Experiment Farm at Kearneysville, and one of the featured speakers, paid his tribute to the 'Golden Delicious' by saying, "Perhaps there have been other areas of greater importance, such as the progress that has been made in the control of our insect and disease problems, but no one single factor has had as great an impact on our industry as the discovery of this one variety of apple" (2).

Unlike the red 'Delicious' which took over a decade to gain renown, the 'Golden Delicious' was immediately acclaimed and soon became a leading cultivar in the United States and abroad. Over 1.5 billion pounds of 'Golden Delicious' are grown annually in the United States, and at least this many are grown in the rest of the world. This multipurpose apple is popular for both fresh and processing uses.

As humbly as the 'Golden Delicious' originated, so it expired in the spring of 1958. Today, all that remains are a few remnants of the steel cage covered by dense underbrush. However, its millions of progeny are scattered throughout the world wherever apples are grown, as monuments to a great West Virginian.

Literature cited

1) Cummins, J.N., P.L. Forsline, and R.D. Way. 1977. A comparison of russeting among 'Golden Delicious' subclones. *HortScience* 12(3):241-242.

2) Terry, D. 1966. *The Stark Story*. Missouri Historical Society. St. Louis, Missouri.

Additional reading

Maas, V. 1970. Golden Delicious. pp. 69-85. In North American apples: varieties, rootstocks, outlook. Michigan State Univ. Press, East Lansing.

Joseph Barrat and David Quinn (WVU) made a pilgrimage to the original Golden Delicious tree in 1958. They were dismayed to find that the tree had died. Prof. Quinn, who is also an artist, made this sketch which appeared on the cover of the 34th Proceedings of the Cumberland-Shenandoah Fruit Workers' Conference.

Authors Baugher and Blizzard were with West Virginia University when this article originally appeared, October 1987.

The 'Kiwifruit'

by I.J. Warrington

Kiwifruit is a comparatively new arrival on the fruit markets of the western world. It is native to southwestern China and, although it was introduced to Europe, the United States, and New Zealand at the beginning of this century, it has only been traded internationally since the mid 1950s.

The kiwifruit (*Actinidia deliciosa* [A. Chev.] C.F. Liang et A.R. Ferguson) is a vine, and fruit are borne on laterals that are produced from one-year-old shoots or canes. Cane renewal is carried out during winter pruning while pruning in summer is used to maintain order in the vines by removing unwanted and tangled growth.

Vines, grown from grafted scions on seedling rootstocks or from rooted cuttings, are trained on many different structures, but the most commonly used are the T-bar fence and the pergola. In common with many other horticultural tree and vine crops, a myriad of alternative structures are being evaluated currently, including modifications of the main growing systems and variations such as V-, Y-, and A-shaped structures.

It is a warm-temperature deciduous species that has a distinct winter dormant phase. Winter chilling is required to ensure adequate budbreak of fruitful laterals in spring. Dormant vines will tolerate mild frosts without injury, but, under New Zealand conditions, temperatures below -12°C are lethal. Flowers and young growth in spring are intolerant of frost. Fertilizer and irrigation requirements are typical of those for similar orchard crops, but shelter from wind is essential to avoid vine and fruit damage.

Kiwifruit is dioecious. Production throughout the world is based almost solely on one pistillate cultivar: 'Hayward.'

Some others, including 'Bruno,' 'Abbott,' and 'Monty,' together with local selections, are grown, but their importance is minor. The main staminate cultivars, which are interplanted in the orchard to supply pollen at flowering time, are 'Matua' and 'Tomuri.' This range of cultivars is likely to increase as current breeding and selection programs identify more precocious pistillate selections with a range of different fruit types, and identify staminate selections with improved pollen production and longer flowering periods.

The success of this fruit is probably attributable to a number of factors. One of the most important is its internal appearance. Cut transversely, the fruit has a creamy-white central core and a brilliant green outer flesh. Light colored bands radiate through the green flesh and small, brown-black seeds surround the core in the spaces between these rays. The skin, which is brown in color, and covered with easily removed hairs, can be eaten but is usually peeled off before the rest of the fruit is consumed. The fruit is very high in vitamin C content (80-120 mg/100 g) and has a pleasant, mild flavor. Fruit are harvested while they are still firm and can be cool-stored for long periods. Ripening is stimulated by ethylene.

Kiwifruit was first introduced into western horticulture at the turn of the century. The origin of 'Hayward' can be traced back to the importation of seed from a Scottish mission in China to New Zealand in 1904. Nurserymen in New Zealand actively promoted its use in home gardens and were quick to recognize the need for male and female plants to be planted together to ensure fruit production. The first small-scale commercial orchards were planted in the 1930s, but it was not until 1952 that the first fruit was exported.

'Hayward' was named after an early New Zealand nurseryman H.R. (Hayward) Wright, who selected the strain originally. The cultivar 'Bruno' was named after another early nurseryman B.H. (Bruno) Just, while 'Allison,' which was raised by Just, was named after Alexander Allison, who probably grew the first kiwifruit plants in New Zealand.

Ironically, seed of kiwifruit was sent to the United States in the same year that seed was sent to New Zealand (1904). Plants were established at the Plant Introduction Station at Chico, California, and these grew successfully and had their first flowers in 1907—unfortunately, they were all staminate or male plants. Pistillate plants were imported from nurserymen in England in 1913, but, in spite of widespread trials, no commercial orchards were established at that time.

Hayward Wright sent seed to Chico in 1935 and plants were successfully established from that introduction. 'Chico' or 'Chico Hayward' was derived from that source. Although this material was established and maintained at the station, it was not promoted and remained overlooked until 25 years later, when the first orchards were established in California in the early 1960s.

Much of the commercial development of kiwifruit, including the selection and naming of cultivars; the development of training, pruning and storage methods; the setting of packaging and grading standards; and the promotion of the fruit to consumers around the world, was, until

Ironically, seed of kiwifruit was sent to the United States in the same year that seed was sent to New Zealand (1904).

very recently, carried out solely by the New Zealand-based industry. In 1986, the area of land planted in this crop in New Zealand was approximately 17,000 hectares and the export volume was approximately 47 million trays (175,000 tonnes).

Although more than half of the volume of fruit presently traded internationally comes from New Zealand, the crop is now grown commercially in many countries in both the Northern and Southern Hemispheres, including the United States (mainly California), France, Italy, Greece, Spain, Japan, Australia, southern Africa, and Chile. By 1990, the total world production of kiwifruit will be about 500,000 tonnes, of which about 60% will be produced in New Zealand.

The kiwifruit is becoming more widely known and accepted by consumers around the world, and it is now regularly traded in many fresh fruit markets. It is likely that new cultivars will soon be developed that reduce the total reliance on 'Hayward.' It is also obvious that current research will develop our knowledge and lead to improvements in the management and handling of this crop both during production and after harvest. The range of processed products available will no doubt be extended as the total volume of fruit produced around the world increases. In particular, kiwifruit production in China will increase, and other *Actinidia* species, with their centers of origin in that country, are likely to be introduced and be developed to become more important in western horticulture.

Author Warrington was with the Plant Physiology Division, DSIR, Palmerston North, New Zealand, when this article originally appeared, January 1988.

References

Ferguson, A.R. 1984. Kiwifruit: A Botanical Review. *Horticultural Reviews* (J. Janick ed.), 6:1-64.

Ferguson, A.R. and Davison, R.M. 1986. *Actinidia deliciosa*. CRC Handbook of Flowering, (A. H. Halevy ed.), Volume V: 1-14.

Ferguson, A.R. and Lay Yee, M. 1983. Kiwifruit (*Actinidia chinensis* var., *hispida*). In: *Plant Breeding in New Zealand* (G.S. Wratt and H.C. Smith eds.). Butterworths, Wellington, New Zealand, 111-16.

Sale, P.R. 1983. Kiwifruit Culture (D. Ashenden Williams ed.) Government Printer, Wellington, New Zealand. 95 pp.

Yerex, D. and Haines, W. 1983. *The Kiwifruit Story*. Agricultural Publishing Associates, Masterton, New Zealand. 96 pp.

The 'Jonathan' Apple

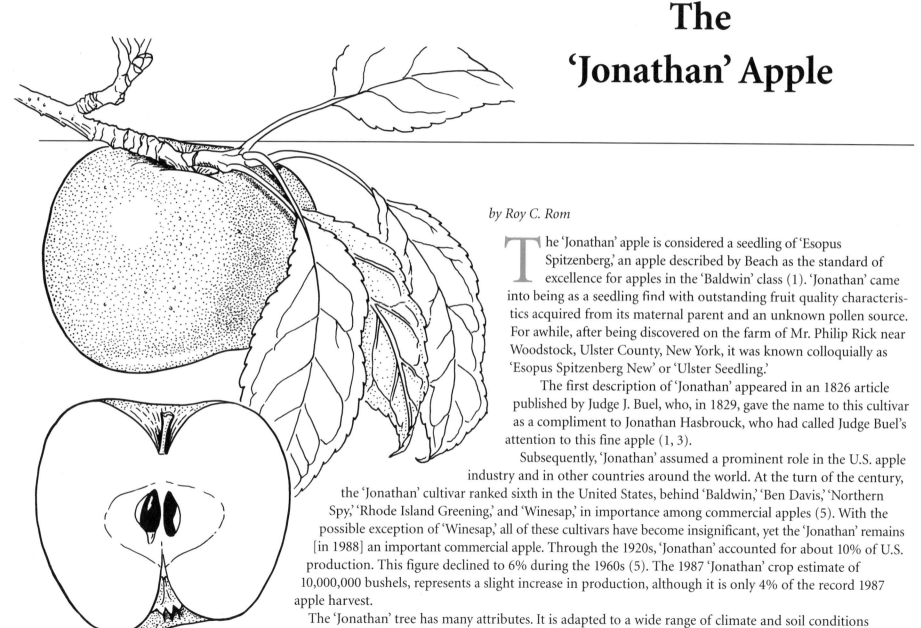

by Roy C. Rom

The 'Jonathan' apple is considered a seedling of 'Esopus Spitzenberg,' an apple described by Beach as the standard of excellence for apples in the 'Baldwin' class (1). 'Jonathan' came into being as a seedling find with outstanding fruit quality characteristics acquired from its maternal parent and an unknown pollen source. For awhile, after being discovered on the farm of Mr. Philip Rick near Woodstock, Ulster County, New York, it was known colloquially as 'Esopus Spitzenberg New' or 'Ulster Seedling.'

The first description of 'Jonathan' appeared in an 1826 article published by Judge J. Buel, who, in 1829, gave the name to this cultivar as a compliment to Jonathan Hasbrouck, who had called Judge Buel's attention to this fine apple (1, 3).

Subsequently, 'Jonathan' assumed a prominent role in the U.S. apple industry and in other countries around the world. At the turn of the century, the 'Jonathan' cultivar ranked sixth in the United States, behind 'Baldwin,' 'Ben Davis,' 'Northern Spy,' 'Rhode Island Greening,' and 'Winesap,' in importance among commercial apples (5). With the possible exception of 'Winesap,' all of these cultivars have become insignificant, yet the 'Jonathan' remains [in 1988] an important commercial apple. Through the 1920s, 'Jonathan' accounted for about 10% of U.S. production. This figure declined to 6% during the 1960s (5). The 1987 'Jonathan' crop estimate of 10,000,000 bushels, represents a slight increase in production, although it is only 4% of the record 1987 apple harvest.

The 'Jonathan' tree has many attributes. It is adapted to a wide range of climate and soil conditions

TABLE 1.

Cultivars with 'Jonathan' as the female (seed) parent:

Cultivar Name	Pollen parent	Year introduced[V]	Harvest season[Z]	Place of origin	Literature source	Germplasm inventory listed[Y]
Akane	Worcester Permain	1953	ahead	Morioka, Japan	1, 6, 21	x
Alps-Cotone	Fugi	?	?	Nagano, Japan	1	
Ancuta	Sugag	?	?	Romania	16	
Agori #1	McIntosh	1953	after	Aomori, Japan	21	
Ardelan	Peasgood	?	?	Romania	16	
Chieftan	Delicious	1966	with	Ames, Iowa, USA	22	x
Choko 64	Terry	1976	after	Nagano, Japan	21	
Delia	Wagener	?	?	Romania	16	
Falticeni	Wagener	?	?	Romania	16	
Frumos de Voinesti	Belle de Boskoop	1980	with	Romania	16, 19	
Fukutami	Ralls Janet	1974	after	Aomori, Japan	10, 21	
Gosho	Northern Spy	1925	with	Aomori, Japan	21	
Hakko	Northern Spy	1925	after	Aomori, Japan	21	
Hatsvaki	Golden Delicious	1976	with	Morioka, Japan	1	
Hirano	Northern Spy	1925	with	Aomori, Japan	21	
Holly	Delicious	1971	after	Wooster, Ohio, USA	2	
Idared	Wagener	1942	just after	Mosion, Idaho, USA	10, 22	x
Jonadel	Delicious	1958	with	Ames, Iowa, USA	10, 22	x
Jonsib	Ikutsk	1938	?	Brookings, SD, USA	22	
Jonwin	Baldwin	1944	ahead	Ettersburg, Calif, USA	22	x
King David	Arkansas Black	1902	after	Durham, Ark, USA	10	x
Lucullus	Cox's Orange	1955	after	Wageningen, Netherlands	10	
Majda	Golden Nobel	1986	just after	Maribor, Yugoslavia	17	
Melrose	Delicious	1944	after	Wooster, Ohio, USA	10, 22	x
Mimi	Cox's Orange	1935	after	Wageningen, Netherlands	10	
Monroe	Rome Beauty	1947	after	Geneva, NY, USA	10, 20, 23	x
Murasaki	Delicious	1948	after	Aomori, Japan	10	
Mutsu[X]	Northern Spy	1939	?	Aomori, Japan	21	
President Boudewijn	Cox's Orange	1952	after	Wageningen, Netherlands	10	
Prime Red (Akane)	Worcester Permain	1970	ahead	Morioka, Japan	1	x
Prins. Bernhard	Cox's Orange	1935	after	Wageningen, Netherlands	10	
Prinses Irene	Cox's Orange	1935	after	Wageningen, Netherlands	10	
Prinses Margriet	Cox's Orange	1955	after	Wageningen, Netherlands	10	
Prinses Marijke	Cox's Orange	1952	with	Wageningen, Netherlands	10	
Rosu de Cluj	Senator	?	after	Romania	16	
Sayaka	Sekai-Ichi	1984	just after	Nagano, Japan	1, 21	
South Dakota Ben	Tony	1938	after	Brookings, SD, USA	22	
South Dakota Bison	(M. Bacatta Yellow Transparent)	1933	?	Brookings, SD, USA	22	
South Dakota Bona	Sylvia	1938	after	Brookings, SD, USA	22	
South Dakota Eda	Tony	1940	after	Brookings, SD, USA	22	
Tsugaruu[W]	Northern Spy	1925	after	Aomori, Japan	21	

[Z] Estimated harvest in relation to Jonathan season.
[Y] North American and European fruit and tree nut germplasm inventory. 1981 USDA Misc. Pub. 1409.
[X] Different from Mutsu, which is a Golden Delicious hybrid.
[W] Different from Tsugaru, which is a Golden Delicious Op seedling.
[V] It was not possible to find dates of introduction for all cultivars.

and is said to succeed wherever grown (3). It is most compatible in Ohio, Missouri, Arkansas, Pennsylvania, Virginia, and the irrigated regions of the West. Fruit sizing and cold hardiness are limiting factors in more northerly regions. The tree is not considered a heavy producer, yet it is known for precocity, annual production, a degree of self-fruitfulness, and reliability of production under adverse conditions (2, 4, 5). Trees, described as moderately vigorous, thrive best on fertile soils and under the best cultural practices. Abundant annual growth, though somewhat thin and frequently growing back into the tree, produces a tree with a dense, rounded canopy (1, 5).

The 'Jonathan' fruit has been definitively described and characterized. Its admirable attributes may be summarized as "beautiful," "brilliant," "having sprightly vinous flavor," and "possessing excellent culinary and dessert qualities" (1, 3).

The 'Jonathan' cultivar is not without weaknesses. The tree is well-known for its susceptibility to fireblight (*Erwinia amylovora* Burr.), both twig and blossom form, cedar apple rust (*Gymnosporanuium juniperiviginanae* schw.), powdery mildew (*Podosphaeda leucotricha* Sal.), bitter rot (*Glomerella cingulata*), and to a lesser extent, apple scab (*Venturia inaequalis* Cke.) (1, 5). These diseases, if not managed effectively, lead to low grade, uneven, irregular, and/or small fruit (4). The plum curculio (*Conutrachelus nensphar* Habst) also has an affinity for the 'Jonathan.' The fruit has several failings that have prevented the 'Jonathan' from becoming a more important commercial cultivar. Fruit sizing is a major problem except when grown under the best management. The fruit is subject to poor finish, principally russet, because the skin is sensitive to early season low temperature and the fungicides needed for disease control (5). A serious fruit shortcoming is that the 'Jonathan' does not store well, although Missouri, 'Jonathans' harvested in 1903 were cited for excellent flavor and appearance when exhibited during August at the 1904 St. Louis Louisiana Purchase Exposition (4). Fruit harvested at an advanced maturity, when overcolor is at its best, is subject to a physiological disorder called Jonathan Spot. This disorder can be checked by holding fruit at temperatures below 1.6°C (35°F); however, soft scald, another physiological disorder, develops whenever 'Jonathans' are stored below 2.2°C (36°F) (6). The development of controlled atmosphere (CA) storage has alleviated the dilemma of these two storage disorders.

The very admirable characteristics of wide geographic adaptation, reliable production, and excellent table and market qualities have resulted in the use of 'Jonathan' as parent of choice by fruit breeders around the world. As a consequence, numerous new cultivars with a strong 'Jonathan' genetic background have been developed and introduced over the past 75 years.

A compilation of information from references and personal communications with apple breeders is presented in the tables. The harvest season, in some instances, is an estimate in relation to a 'Jonathan' harvest. The necessity to make an estimate points to the fact that cultivar descriptions notably lack this information or do

TABLE 2.
Cultivars with 'Jonathan' as the pollen parent:

Cultivar Name	Female seed parent	Year introduced[V]	Harvest season[Z]	Place of origin	Literature source	Germplasm inventory listed[Y]
Appel van Paris	Barbant	1935?	after	Wageningen, Netherlands	10	
Aromat de Vara	Parmain D'or	?	before	Romania	16	
Conrad	Ben Davis	1935	just after	Mt. Grove, Mo, USA	22	x
Crandall	Rome Beauty	1952	?	Urbana, Ill, USA	22	x
Directeor van de Plassche	Cox's Orange	?	before	Wageningen, Netherlands	10	
Faurot	Ben Davis	1935	after	Mt. Grove, Mo, USA	22	x
Florina (Querina)	(complex hybrid)	1977	after	Angers, France	8, 18	
Fyan	Ben Davis	1920	after	Mt. Grove, Mo, USA	22	x
Himekami	Fuji	1984	with	Morioka, Japan	1, 9	
Holiday	Macoun	1964	after	Wooster, Ohio, USA	10, 22	x
Idajon	Wagener	1949	before	Moscow, Idaho, USA	10, 22	x
Iwakami	Fuji	1984	with	Morioka, Japan	9, 21	
Joan	Anisim	1932	?	Ames, Iowa, USA	22	x
Jonagold	Golden Delicious	1968	after	Geneva, NY, USA	13, 22, 23	x
Jonagram	Ingram	1956	with	Mt. Grove, Mo, USA	22	x
Jonalicious	seedling, unknown parentoue	1960	with	Abeline, Tex, USA	14, 22	x
Jonamac	McIntosh	1972	before	Geneva, NY, USA	11	x
Jono	Summer Champion	1972	before	Stillwater, Okla, USA	3	x
King Cole	Dutch Mignone (possible reverse cross)	1912?	after	Victoria, Australia	10	x
Kogetsu	Golden Delicious	1968	before	Aomori, Japan	1, 9, 21	
Lonjon	London Pippin	1975	before	Maribor, Yugoslavia	17	
Malling Kent[X]	Cox's Orange	?	?	East Malling, England	15	x
Megumi	Ralls Janet	1950	after	Aomori, Japan	10, 21	
Minjon	Wealthy?	1942	with	Excelsior, Minn, USA	10, 22	x
Priam	Pri 14-126	1974	with	NJ, Ind, Ill, USA	4	x
Prinses Beatrix	Cox's Orange	1935?	after	Wageningen, Netherlands	10	
Red Granny Smith	Granny Smith	1961?	after	West Australia	10	
Rensselar	Ben Davis	1914	?	Geneva, NY, USA	20	
Secor	Salome	1922	after	Ames, Iowa, USA	22	x
Shinko	Ralls Janet	1948	just after	Aomori, Japan	10, 21	x
Sumukoni	Ralls Janet	?	after	Aomori, Japan	21	
Tamahime	Foster	1940	before	Ehime, Japan	21	
Wamdesa	Elk River	1938	?	Brookings, SD, USA	22	
Webster	(Ben Davis x Jon) x (Ben Davis x Jon)	1935	after	Geneva, NY, USA	20, 22	x
Wright	Ben Davis	1942	same	Mt. Grove, Mo, USA	22	x

[Z] Estimated harvest in relation to Jonathan season.
[Y] North American and European fruit and tree nut germplasm inventory. 1981, USDA Misc. Pub. 1406.
[X] Jonathan background suspected but not confirmed.
[W] It was not possible to find dates of introduction for all cultivars.

TABLE 3.

Cultivars originating as mutations of 'Jonathan':

Cultivar Name	Year introduced[V]	Harvest season[Z]	Place of origin	Literature source	Germplasm inventory listed[Y]
Allred	1937	with	Allentown, Penn, USA	14	
Anderson Jonathan	1927	with	Covert, Minn, USA	14	x
Black Jon	1931	with	Wenatchee, WA, USA	10, 14	x
Conkle Jonathan	1943	with	Chester, WV, USA	10	x
Edwards Jonathan	1949	with	Quincy, Ill, USA	22	
Erlison[X]	1968	ahead[Z]	Green Forest, Ark, USA	22	
Jon-a-red	1934	with	Peshastin, Wash, USA	10, 14	x
Jonnee	1967	with	Cladwell, Idaho, USA	14, 22	x
Jonathan-Apex	1961	with	Kelowna, BC, Canada	10	
Kapai Red Jonathan	1931	with	Hawkes Bay, New Zealand	10	
KingJon (King Red Jonathan)	1933	with	Wenatchee, Wash, USA	22	x
Nu-Jon[X]	1949	ahead	Entiat, Wash, USA	22	
Nured	1965	with	Wenatchee, Wash, USA	14, 22	x
Purdue Black-Jon	1940	?	Lafayette, Ind, USA	22	
Red Jonathan	1951	with	New Zealand	10	
Valnur (KingJon)	1933	with	Wenatchee, Wash, USA	14	
Watson Jonathan	1950	with	Vernon, BC, Canada	10, 22	x
Welday	?	?	Smith Field, Ohio, USA	22	x
Young Bearing Jonathan	1932	ahead	Vera, Mo, USA	22	

[Z]Estimated harvest in relation to Jonathan season.
[Y]North American and European fruit and tree nut germplasm inventory 1981, USDA Misc Pub.140e.
[X]Some question of a Jonathan mutation.
[W]It was not possible to find dates of introduction for all cultivars.

TABLE 4.

Cultivars containing 'Jonathan' in their genetic background:

Cultivar Name	Parentage	Year introduced[V]	Harvest season[Z]	Place of origin	Literature source	Germplasm inventory listed[Y]
Burgandy	(Jon x Rome) x (Macoun x Antonovka)	1974	ahead	Geneva, NY, USA	12	x
Generos	((Parmin D'ar x (M.Kaido x Jon) x (Jon x Belle de Boskoop)	?	late	Romania	16, 19	
Gloria	Jonathan x (Jonathan x Pasgood) x (Gustav Dudobil x Van Mors)	?	after	Romania	16	
Jonafree	(Jonathan x 14-6 44) x (Gallia Beauty x Redspy)	1979	with	NJ, IN, IL, USA	7	x
Maefree	McIntosh x PRI 48-177 PRI 48-177 has some Jonathan background	1974	ahead	Trenton, Ontario, Canada	5	x
Pionier	(Verzisoare x Jonathan) x Prima	?	just after	Romania	16, 19	
Spijon	Redspy x Seedling Jonathan	1968	after	Geneva, NY, USA	13	x
Voinea	(Verisoare x Jonathan) x Prima	?	?	Romania	16, 19	

[Z]Estimated harvest in relation to Jonathan season.
[Y]North American and European fruit and tree germplasm inventory 1981, USDA Misc Pub 1406.
[X]It was not possible to find dates of introduction for all cultivars.

not reference the harvest time to a standard cultivar. Tables 1 and 2 list named cultivars in which the 'Jonathan' served as seed or pollen parent. Table 3 lists the 'Jonathan' mutations, selected primarily for improved or early color development. Cultivars noted in Table 4 are newer hybrids with 'Jonathan' background.

Obviously, not all the cultivars have become apples of note; many have only local acceptance. Some have not survived the test of time, and others have not been fully assessed. What is important and evident is that the 'Jonathan,' which carries 'Spitzenberg' quality and excels its parent (1), is a revered cultivar. This is confirmed by the fact that it has been so widely used in new apple cultivar development.

'Jonathan' synonyms include, 'Djonathan,' 'Dzhonathan,' 'Dzonetr,' 'Esopus,' 'Spitzenberg (new),' 'King Philip,' 'King Phillip,' 'Philipp Rick,' 'Philip Rick,' 'Pomme Jonathan,' 'Ulster,' and 'Ulster Seedling.'

Literature cited

1) Beach, S.A. 1905. The Apples of New York. Rpt of NY Agr. Exp. Sta. 1903. Vol. 1. John Wiley & Son, NY.

2) Dearing, C.T. (?) The Twelve Most Popular Varieties of Missouri Apples. MO State Board of Hort. Bul. N. 17.

3) Dowing, A.J. 1876. *The Fruit and Fruit Trees of America.* John Wiley and Sons, NY.

4) Gould, H.P. and W.F. Fletcher. 1913. Apples and Peaches of the Ozark Region. USDA Bureau of Plant Industry Bul. 275.

5) Larsen, R.P. 1970. Jonathan. Ch. 6, In: *North American Apples: Varieties, Rootstock Outlook.* Mich State Univ. Press, East Lansing, MI.

6) Porritt, S.W., M. Meheriuk, and P.D. Lidster. 1982. Postharvest Disorders of Apples and Pears. Agr. Canada Pub. 1737 IE.

Literature source for cultivar descriptions

1) *Fruit Varieties Journal.* 1987. 41(1):22-25.

2) *HortScience.* 1971. 6(5):439.

3) *HortScience.* 1972. 7(5):457.

4) *HortScience.* 1974. 9(4):401.

5) *HortScience.* 1975. 10(5):472.

6) *HortScience.* 1978. 13(5):522.

7. *HortScience.* 1979. 14(4):551.

8) *La France Agricole.* Dec. 1985.

9) Nakajima—Tenkoen Co. LTD Catalogue 1987. Higashin City, Japan.

10) *National Apple Register of the United Kingdom.* 1971.

11) New York Food and Life Sciences Bul. 25. 1972.

12) New York Food and Life Sciences Bul. 47. 1974.

13) New York State Ag. Exp. Sta. Circ.12, 1968.

14) *North American Apples: varieties, rootstocks, outlook.* 1970. Michigan State University Press.

Personal communications 1987

15) Dr. F.H. Alston. Inst. Hort. Res., East Malling, England.

16) Dr. V. Cociu. Research Institute of Pomology, Maracinena, Romania.

17) Dr. J. Cyrnko. Biotehniska Fakulteta, lzpostava. Maribor, Yugoslavia.

18) Dr. M. Le Lezek. Station De Recherches D'Arboriiviture, Beaucouze, France.

19) Dr. L. Serboiu. Voinesti Fruit Res. Sta. Voinesti, Romania.

20) Dr. R. Way. NY State Agri. Exp. Sta. Geneva, New York.

21) Dr. Yoshiro Yoshida. Fruit Tree Res. Sta., Yatabe, Japan.

22) *Register of New Fruit and Nut Varieties,* 2nd ed. 1972. University of California Press.

23) *Search Agriculture* 1971. NYSAES V. 1, V. 2., Geneva, NY.

Admirable attributes of 'Jonathan' may be summarized as "beautiful," "brilliant," "having sprightly vinous flavor," and "possessing excellent culinary and dessert qualities."

Author Rom was with the Department of Horticulture and Forestry, University of Arkansas, Fayetteville, when this article originally appeared, April 1988.

The 'Montmorency' Sour Cherry

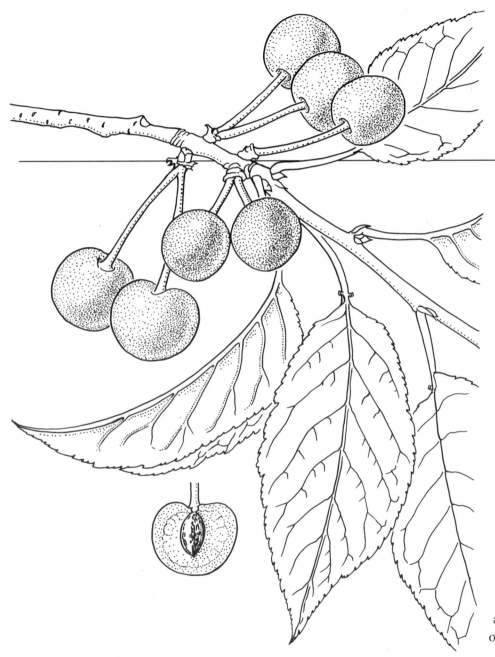

by Amy F. Iezzoni

'Montmorency' reigns supreme as the only sour cherry cultivar recommended for commercial planting in the United States; it currently represents approximately 99% of the sour cherry production. Within the United States, 'Montmorency' production is concentrated in those areas where the mean summer temperature (June, July, and August) is near 15°C (5), the minimum winter temperature is -35°C (G. S. Howell, personal communication), and spring freeze damage is limited. As a result, approximately 75% of the U.S. crop is produced in Michigan. Next in order of importance are New York, Utah, Wisconsin, and Pennsylvania, with increasing acreages in Washington and Oregon. 'Montmorency's' success is its productivity despite low crop years due to occasional spring freezes, and its fruit quality, which has become the industry standard. The skin of the fruit is bright red, the juice clear, the pit relatively round, and the flesh fairly firm.

'Montmorency's' growth habit is upright, and therefore, careful training is required to obtain trees with strong leaders and wide branch angles which resist breakage. Once the tree is trained, regular pruning is recommended to increase light penetration into the canopy and thereby keep fruit spurs in production. On older trees, an increasing proportion of the fruit buds are produced laterally on one-year-old wood.

'Montmorency' fruit is harvested when the soluble solids content is at least 11% and the fruit diameter is approximately two centimeters. Mechanical trunk shakers harvest almost the entire sour cherry acreage in the United States. Once harvested, the fruit is cooled, washed, sorted mechanically, pitted, and processed into pie filling, or frozen for later use. During processing, sugar and red colorants are frequently added.

'Montmorency' probably originated before the seventeenth century in the Montmorency Valley in France as an open pollinated seedling of 'Cerise Hative' or 'Cerise Commune' (4); this would make it almost 400 years old! The date of importation into the United States is unknown; however, the date of the first written reference to 'Montmorency' is 1832. The American Pomological Society added 'Montmorency' to its fruit catalogue in 1897, using the qualifying term 'Ordinaire;' this was dropped in 1909. Numerous spontaneous mutations for ripening date, fruiting habit, and fruit size have been identified (2); however, none has yielded significantly more than 'Montmorency' (3). 'Montmorency' has been used as a parent in breeding programs in Ontario, New York, and Minnesota, but only one cultivar, 'Meteor,' was released from these breeding efforts. 'Meteor,' resulting from a cross between 'Montmorency' and the Russian cultivar 'Vladimir,' was released from the University of Minnesota in 1952 (1). However, since 'Meteor's' pit is oblong, the pit fragments when the fruit is mechanically pitted, making the fruit from this cultivar commercially unacceptable.

In contrast to its popularity in the United States, 'Montmorency' accounts for only 20% of the sour cherry production in the Northern Hemisphere. It is rarely grown in Europe and the Soviet Union, because sour cherries with red to dark red juice are preferred by consumers, and 'Montmorency' is unproductive under their environmental conditions. Researchers from Yugoslavia (7) to Norway (6) have classified 'Montmorency' as only partially self-fertile in comparison with their more productive cultivars. In addition, 'Montmorency' is extremely susceptible to leaf spot caused by *Coccomyces hiemalis*, and new cultivars with a high level of tolerance to this fungus are being grown in the Soviet Union, Yugoslavia, Romania, and Hungary.

In the United States, 'Montmorency' is also not without its shortcomings. Sour cherry yields in the Great Lakes states are frequently reduced one out of every three years by spring freezes. 'Montmorency' is also susceptible to numerous fungal and virus diseases. In certain years, the fruit may be soft and juice is lost during processing.

'Montmorency' has served the U.S. sour cherry industry well, but the need for market expansion through product diversification, the prospective loss of chemical sprays, and competition with other new cultivars may necessitate finding replacements for this 400-year-old cultivar. Hopefully, these will be available for 'Montmorency's' 500th birthday.

> *'Montmorency' probably originated before the seventeenth century in the Montmorency Valley in France.*

Literature cited

1) Alderman, W.H. et al. 1952. Meteor cherry. Minn. Agric. Exp. Stn. Misc. Rep. 16.

2) Drain, B.D. 1933. Field studies of bud sports in Michigan tree fruits. Mich. State Coll. Tech. Bull. 130.

3) Goldy, R.G., R.L. Andersen, and F.G. Dennis, Jr. 1982. Phenotypic and cytologic analysis of spontaneous mutations of the 'Montmorency' cherry (*Prunus cerasus* L.). *J. Am. Soc. Hortic. Sci.* 107:779-781.

4) Hedrick, U. P. 1915. *The cherries of New York.* J. B. Lyon. Albany, NY.

5) Marshall R.E. 1954. *Cherries and cherry products.* New York Interscience Publishers, Inc.

6) Redalen, G. 1984. *Fertility in sour cherries.* Gartenbauwessenschaft 49:212-217.

7) Stancevic, A.S. 1975. Self-fertility of sour cherry trees. *J. Yugoslav Pomology* 31-32.

Author Iezzoni was with the Department of Horticulture, Michigan State University, East Lansing, when this article originally appeared, July 1988.

The 'Stuart' Pecan

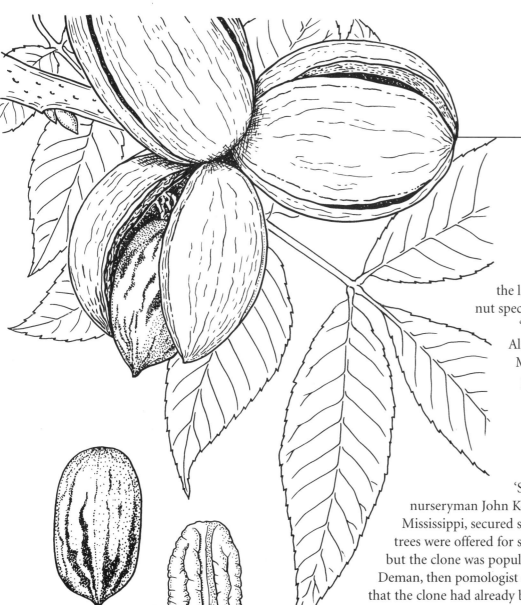

by Tommy E. Thompson

The most popular pecan cultivar (variety) in the world is 'Stuart,' currently occupying about 27% of improved orchard space (6). Worldwide, the United States is by far the largest producer of this most important native North American nut species.

'Stuart' presumably originated from a nut brought from Mobile, Alabama, by John R. Lassabe and planted in a garden in Pascagoula, Mississippi, about 1874 (4). The tree quickly acquired local popularity due to its productiveness, beauty, and nut quality compared to contemporary trees. From 1889 to 1892, it annually produced about 140 pounds of nuts, and in 1892, it yielded 350 pounds which were sold by Charles M. Cruzat for $1 per pound. Cruzat had the land leased where the 'Stuart' tree was growing.

Mr. A.G. Delmas of Scranton, Mississippi, first cut scions of 'Stuart' in 1886 and got one of 60 grafts to take (4). In 1890, nurseryman John Keller, associated with Col. W.R. Stuart of Ocean Springs, Mississippi, secured scions from the tree and propagated nursery trees. The first 'Stuart' trees were offered for sale about 1892. The original name for this cultivar was 'Castanera,' but the clone was popularized under the name 'Stuart,' suggested by Prof. H.E. Van Deman, then pomologist of the U.S. Department of Agriculture. Van Deman was unaware that the clone had already been named. 'Stuart' was widely advertised and sold throughout the South.

> *If 'Stuart' is a parent of 'Schley,' and 'Schley' a parent of 'Mahan,' then 'Stuart' is a major contributor of genes for our modern USDA breeding program.*

In October 1893, the original tree in Captain Castanera's garden was blown down by a storm. The tree sprouted and again bore nuts in 1902, but another storm blew the tree down in 1906, and it died.

Whether 'Schley' and thus other more recently developed cultivars are descendants of this clone is currently receiving some debate. 'Schley' is reported to be a seedling of 'Stuart' (5), but if 'Stuart' was planted in 1874 and 'Schley' in 1881, 'Stuart' would have had to produce nuts in its seventh year. This is possible, but is discounted (1) since 'Stuart' is so lacking in precocity. Perhaps 'Stuart' was planted before 1874, since the literature says it was planted "about 1874."

Proceeding forward, 'Mahan' is possibly a 'Schley' seedling. If 'Stuart' is a parent of 'Schley,' and 'Schley' a parent of 'Mahan,' then 'Stuart' is a major contributor of genes for our modern USDA breeding program. It should also be noted that the 'Mahan'-'Schley' lineage is supported by recent isozyme inheritance research (2). More recent attempts to use 'Stuart' as a direct parent to produce modern cultivars have been very disappointing, however.

'Stuart' trees have a strong upright type of growth, with major limbs connected to the trunk quite low and a distinct absence of a main central leader. Limb connections are very strong, and if limb breakage is induced by storms, etc., it usually occurs higher in the tree.

'Stuart' evidently has a high bud chilling requirement since bud break is delayed somewhat after mild winters in southern areas (3). As far as blooming habit, it is strongly protogynous, and a pollinator is needed for all orchards.

Generally, 'Stuart' is very low in precocity, but yields well and fairly regularly after tree maturity (ten years or so). It has a nut thinning mechanism which prevents overbearing to some extent, and contributes to uniformity of yield from year to year. An average yield of 1,200 to 1,500 pounds per acre in well-managed orchards is common.

'Stuart' nuts mature midseason, which is about October 15 at Brownwood and about a week earlier in Georgia. The nut is oblong, slightly compressed (flattened), and has a blunt apex and rounded base (7). Color is brownish gray with somewhat irregular purplish black streaks from the apex part way down.

The nut is of good size (45-55 per pound) compared to other cultivars but has a thick shell and low percent kernel (45-50). Both dorsal and ventral kernel grooves are deep. The dorsal grooves are also wide, usually allowing adequate separation from packing material developed in these grooves. The ventral groove is narrow and more likely to retain a small portion of packing material during shelling. Good 'Stuarts' can shell nicely, producing a high percentage of intact halves.

'Stuart' has moderately high pecan scab resistance, which is of considerable economic value in the more humid areas where it is grown. It is somewhat susceptible to downy spot and vein spot. Although insect resistance is not generally an economic factor in evaluating pecan cultivars, 'Stuart' is quite susceptible to stem phylloxera and shuckworm. This grand old cultivar remains more common in today's pecan orchards than any other [in 1988], and is still recommended for planting in nine states. Due to the almost permanent production potential of pecan trees, 'Stuart' will surely be a popular cultivar for years to come.

Literature cited

1) Anonymous. 1970. This is centennial year for the Stuart. Pecan Q. 4(4):8.

2) Marquard, R.D. 1987. Isozyme inheritance, polymorphism, and stability of malate dehydrogenase and phosphoglucose isomerase in pecan. *J. Amer. Soc. Hort. Sci.* 112(4):717-721.

3) Sparks, D. 1977. Notes on the Stuart pecan. Pecan S. 4(5):204-207.

4) Taylor, W.A. 1905. Promising new fruits. In: *1904 Yearbook of the U.S. Department of Agriculture*, pp. 399-416.

5) Taylor, W.A. 1906. Promising new fruits. In: *1905 Yearbook of the U.S. Department of Agriculture*, pp. 495-510.

6) Thompson, T.E. 1984. Pecan cultivars: current use and recommendations. *Fruit Var. J.* 38(3):103-112.

7) Thompson, T.E. and E.F. Young, Jr. 1985. Pecan cultivars—past and present. Tex. Pecan Grow. Assn., College Station. 265 pp.

Author Thompson was with the United States Department of Agriculture, Agricultural Research Service, Brownwood, Texas, when this article originally appeared, October 1988.

The 'York Imperial' Apple

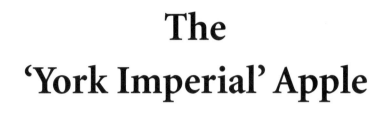

by Howard A. Rollins, Jr.

For many years the 'York Imperial' cultivar dominated the apple industry of the concentrated Appalachian fruit production area extending from northern Virginia to southern Pennsylvania. However, more recently, it has given way to 'Red Delicious' and 'Golden Delicious' cultivars. While 'York Imperial' is primarily a processing apple, improved red strains are finding gradual increased popularity for fresh, late-season markets.

'York Imperial' dates back to the early 1800s when it was found on the farm of Mr. Johnson which adjoined the borough of York, Pennsylvania. It attracted attention because of its long keeping quality. By the early 1900s, it had become widely planted throughout the region for export to European markets, which were quite profitable until about 1930, when British import restrictions sharply changed the Appalachian area markets. It became difficult for 'York Imperial' apples to compete on domestic markets with red cultivars where only very top grades were in demand. The expanding apple processing firms throughout the area utilized lower grades of fruit, but they were largely salvage-type operations.

As apple processors broadened their product lines, and markets for sauce and slices improved, the 'York Imperial' cultivar became preferred by processors over other cultivars. It is a firm-to-hard apple with a creamy yellow flesh, which provides a desirable colored sauce. The firm fruit texture provided slices that held their shape and produced a canned product favored by pie bakers.

The superior keeping quality of the raw product, and its resistance to bruising, favors the handling and holding of fruit for processing. 'York Imperial' is characterized by a lopsided shape and is typically flattened with a small core. When peeled, cored, and trimmed, it results in high yields of processed product from a given weight of raw fruit.

'York Imperial' is a late-season cultivar maturing between 'Stayman' and 'Winesap.' Harvest typically can be extended over several weeks. Currently, processors divide the delivered fruit into three categories—one held in the yard for early processing, a second placed in conventional storage, and a third placed in controlled atmosphere storage for later use to extend the processing season.

'York Imperial' is a high-yielding cultivar and typically requires chemical thinning. Favorable thinning results can be achieved with either naphthalene acetic acid or naphthylacetamide. It does tend to alternate bearing unless crop loads are reduced. If set is not reduced, fruit tends to "rope" up in clusters, and as fruit size increases, individual fruits are forced off, since stems are short.

'York Imperial' is moderately tolerant of the more common apple diseases in the Appalachian region (apple scab and powdery mildew); however, it is quite susceptible to cedar rusts and fireblight. One of the more serious problems is the susceptibility to the physiological disorder 'York' spot or cork spot. This is characterized by one-quarter inch or larger corky areas near the surface of the fruit. The use of soluble boron sprays during bloom or early postbloom periods will reduce but not eliminate the problem.

The 'York Imperial' tree is typically large with an upright growth habit. It is a relatively easy tree to grow and maintain. It is common practice in mature, bearing 'York' trees to make only a few large pruning cuts each year and only limited detailed pruning cuts.

The 'York Imperial' is still a major cultivar throughout the Appalachian fruit region [in 1989]; however, the average age of bearing trees is increasing, and newer orchards are being more widely planted to fresh-market cultivars. As older orchards are eliminated, the total production of 'York Imperial' is expected to decline. Where newer plantings of 'York Imperial' are being established, the trend is toward smaller than standard trees, more closely spaced and intensely managed. Prices received for fruit for processing will play a significant role in new planting trends, and where 'York' trees are planted, improved red coloring strains such as 'Commander York' and 'Red Yorking' will likely dominate.

Additional information can be found in Volume 1 of *Apples of New York* by S.A. Beach and *North American Apples: Varieties, Rootstocks, Outlook,* published by Michigan State University Press East Lansing, 1970.

> *'York Imperial' first attracted attention because of its long keeping quality.*

Author Rollins was with the Virginia Polytechnic Institute and State University, Winchester Agricultural Experiment Station, Winchester, Virginia, when this article originally appeared, January 1986.

The 'Willamette' Red Raspberry

by Hugh A. Daubeny, F.J. Lawrence, & G.R. McGregor

'Willamette' has been the predominant red raspberry cultivar in the Pacific Northwest for almost 40 years [in 1989]. Only in the past few years has there been extensive planting of other cultivars which yield more, have larger fruit sizes, higher levels of pest resistance, and/or better adaptation to fresh and other specialty markets. Since the Pacific Northwest accounts for more than 90% of the total red raspberry production in North America, 'Willamette' can be considered the continent's most important raspberry cultivar. Moreover, outside the Pacific Northwest, it has been grown extensively by the burgeoning industry in central coastal California, and also has been grown in protected locations in eastern North America. In the Southern Hemisphere, it has been an important cultivar in Australia, New Zealand, and Chile. It also has been grown successfully in several eastern European countries including Yugoslavia, where it is now of major importance, and Hungary. By virtue of all of the above, 'Willamette' has the unique distinction of having been and continuing to be the world's most important red raspberry cultivar.

There are interrelated reasons why 'Willamette' has had such immense and enduring success. Among the fruiting traits are the dark purple-red color and the relative ease of removal or abscission. The color and also the texture were ideally suited to the needs of the canning industry, which was still important in the years immediately following the cultivar's release. In more recent years, this color has suited the needs of the rapidly expanding juice industry. The fruit is well presented, and this, combined with the ease of removal or picking, has facilitated hand harvest.

The ease of removal is also ideally suited to harvesting by machines, which are now used extensively in the Pacific Northwest for fruit destined for processing.

'Willamette' has resistance or tolerance to a number of pests, and this, too, has contributed much to its success and longevity. For example, it is resistant to cane Botrytis (*Botrytis cinerea* Pers. ex. Fr.) (7), cane spot *(Elsinoe veneta* Burkh.) (9), powdery mildew (*Sphaerotheca macularis* (Wallr. Fr.) Lind) (3), and to crown gall *(Agrobacterium tumofaciens* [Smith and Towns] Conn) (7). It has also shown some resistance to the black vine weevil, *Otiorhynchus sulcatus* (F.) (2). Despite its susceptibility to *Amphorophora agathonica* Hottes, the aphid vector of the raspberry mosaic virus (RMV) complex, it has been assumed that 'Willamette' is tolerant to the complex since there have been no obvious effects of infection (5). Nevertheless, a recent study has indicated that 'Willamette' can become infected with leaf spot virus (LSV), one of the components of the complex (7), but there is no direct evidence that LSV by itself has an adverse effect on growth and fruiting. The cultivar does not become infected with the strain of the pollen-transmitted virus raspberry bushy dwarf (RBDV) (6).

Subsequently, it does not show crumbly fruit and/or reductions in vigor and yield commonly found in susceptible cultivars that become infected with the virus either alone or in combination with one or more other viruses. Although the cultivar will become infected with the nematode-transmitted virus, tomato ringspot, it usually is not severely affected by it (8). 'Willamette' supports high populations of the root lesion nematode (*Pratylenchus penetrans* (Cobb) Fil. & S. Skek.), but there is some evidence that tolerance to the pest could exist (1, 11).

The plant habit of 'Willamette' allows for relatively easy management compared to many other cultivars. Canes show minimum amounts of branching, are moderately vigorous and have medium-length laterals which tend to grow upright. Production of abundant primocanes has meant that the cultivar has been easy to propagate and also that cane numbers have been easy to manipulate by chemical burning.

Under optimum growing conditions, 'Willamette' plants will establish and grow more rapidly than those of some other cultivars. Thus, the cultivar often produces a substantial crop the year after planting.

'Willamette' has remained genetically stable compared to some other cultivars. Of particular importance is the fact that mutations resulting in reduced fertility, expressed as crumbly fruit, have rarely been recorded.

In warmer climates, such as the production areas of southeastern Australia and central coastal California, 'Willamette' shows a greater level of heat tolerance than many other cultivars. In addition, it will produce a fall crop on the primocanes in these climates. This crop, although later than that of the standard primocane fruiting cultivar 'Heritage,' is of high quality. Alternative year cropping offers the option of manipulating the level of primocane fruiting, which makes 'Willamette' a particularly flexible cultivar.

'Willamette' grows well in a range of soil

> *'Willamette' is the world's most important red raspberry cultivar.*

types and has at least a partial degree of drought tolerance. It is not suited to heavier soils which might have drainage problems during the winter months. Under such conditions, the cultivar often shows root rot (most likely caused by *Phytophthora erythroseptica* Pethybr.) (7).

'Willamette' is also susceptible to several other diseases, including spur blight (*Didymella applanta* (Niessl) Sacc.) (4) and bacterial blight (*Pseudomonas syringae* van Hall) (10). The fruit is relatively susceptible to both pre- and post-harvest fruit rots; the former is mostly caused by *B. cinerea* and the latter by both *B. cinerea* and *Rhisopus* spp. (probably *R. stolonifer* [Ehr. ex. Fr.] Vuill.) (7). Susceptibility to postharvest fruit rots, along with the dark purple-red color and lack of a high level of firmness, limits the shelf life of the fruit. This limitation, plus somewhat acidic flavor, has meant the fruit has never been considered particularly well suited to the fresh market.

'Willamette' was selected in 1936 from the cross of 'Lloyd George' x 'Newburgh,' made in 1933 by George Waldo working at Corvallis, Oregon, in the Oregon State University-United States Department of Agriculture breeding program there. It was named in 1943, and within 10 years, dominated the Pacific Northwest red raspberry industry. It has been used extensively in breeding programs. To date, 'Meeker' from the Washington State University breeding program at Puyallup, is the most notable cultivar which has 'Willamette' as a parent. It is expected that 'Willamette' and cultivars derived from it will continue to have major effects on worldwide red raspberry production for many years to come.

Author Daubeny was with Agriculture Canada, Vancouver, British Columbia; Lawrence with the National Clonal Germplasm Repository, Corvallis, Oregon; and McGregor with the Potato Research Station, Healesville, Victoria, Australia, when this article first appeared, April 1989.

Literature cited

1) Bristow, P.R., B.H. Barritt, and F.D. McElroy, 1980. Reaction of red raspberry clones to the root lesion nematode. *Acta Hort.* 112:39-43.

2) Cram, W.T. and H.A. Daubeny. 1982. Responses of black vine weevil adults fed foliage from genotypes of strawberry, red raspberry, and red raspberry-blackberry hybrids. *HortScience* 17:771-773.

3) Daubeny, H.A., P.B. Topham, and D.L. Jennings. 1968. A comparison of methods for analyzing inheritance data for resistance to red raspberry powdery mildew. *Can. J. Genet. Cytol.* 10:341-350.

4) Daubeny, H.A. and H.S. Pepin. 1974. Susceptibility variations to spur blight *(Didymella applanata)* among red raspberry cultivars and selections. *Plant Dis. Reptr.* 58: 1024-1027.

5) Daubeny, H.A.1980. Red raspberry cultivar development in British Columbia with special reference to pest response and germplasm exploitation. *Acta Hort.* 112:59-67.

6) Daubeny, H.A., J.A. Freeman, and R. Stace-Smith. 1982. Effects of raspberry bushy dwarf virus on yield and cane growth in susceptible red raspberry cultivars. *HortScience* 17:645-647.

7) Daubeny, H.A. 1986. The British Columbia raspberry breeding program since 1980. *Acta Hort.* 183:47-58.

8) Freeman, J.A., and R. Stace-Smith. 1968. Effects of tomato ringspot virus on plant and fruit development of raspberries. *Can. J. Plant Sci.* 48:25-29.

9) Jennings, D.L., and G.R. McGregor. 1988. Resistance to cane spot *(Elsinoe veneta)* in the red raspberry and its relationship to resistance to yellow rust *(Phragmidium rubi-idaei)*. Euphytica 37:173-180.

10) Pepin, H.S., H.A. Daubeny and I.C. Carne. 1967. Pseudomonas blight of raspberry. *Phytopathology* 57:929-931.

11) Vrain, T.C., and H.A. Daubeny. 1986. Relative resistance of red raspberry and related genotypes to the root lesion nematode. *HortScience* 21:1435-1437.

The 'Barcelona' Hazelnut

by Shawn A. Mehlenbacher & Anita N. Miller

'Barcelona' is the leading cultivar of the European hazelnut (*Corylus avellana* L.) grown in Oregon. Members of this species are known by many names owing to the great diversity in size and shape of the nuts, husk length, and geographic origin, and the rich folklore associated with it. Synonyms include filbert, lambert nut, and Pontic nut.

The term *hazel* is derived from the Anglo-Saxon *haesel*, a hood or bonnet. Variations of the term are used throughout northern Europe. The word "filbert" is believed to be a corruption of the English "full-bearded" (5), referring to the long, fringed husks which extend past and enclose the nuts. For centuries, the short-husked types were called hazelnuts and the long-husked types were called filberts.

Another possible origin of the word "filbert" was pointed out by Bunyard (1920). He noted that it may not be mere coincidence that St. Philibert's day falls on August 22, coinciding with the ripening of the nuts.

Hazelnut production in the Italian province of Campania has been important in the village of Avellino for centuries. The Spanish word for hazelnut is avellano. These two, and the species name adopted by Linnaeus, are believed to be derived from Abeliana, the location in Asia from which the nuts were supposedly introduced to Europe.

The term "lambert nut" is thought to be a corruption of "long-bearded" or the German equivalent *Langbart*. Others attribute it to the Italian Lombard family who introduced the nut to Britain. The term "Pontic nut" is used for hazelnuts from Turkey, where they are grown extensively between the Pontic mountains and the Black Sea coast. Additional synonyms are discussed by Lagerstedt (1975).

In the early years of hazelnut production in Oregon, growers adopted the term "filbert" to distinguish their superior product from the native wild "hazelnuts." However, in order to improve marketing, there has been a recent shift to using hazelnut, the more universally recognized name.

History

'Barcelona' and indeed, the entire hazelnut industry in the Pacific Northwest, can be traced to one man, Felix Gillet (1835-1908). Born in Roucheford, France, he emigrated to the United States in 1852. During a two-year stay in Boston, he learned the barber's trade, and in 1859, he moved and set up shop in Nevada City, California. Although barbering remained his full-time occupation until 1870, his interests shifted to the plants he knew as a boy. In 1864, he returned to France for 10 months to thoroughly learn the nursery business. By 1870, he had saved enough money to buy a 16-acre piece of rocky, barren land on the outskirts of Nevada City. Although he continued to work as a barber, his spare time was devoted to clearing and tilling, as one day he expected to own and operate a nursery which would command attention (1).

His neighbors predicted failure. Fairchild (1938) describes the subsequent events as follows: "Then, with the confidence of a man who knows that he can grow plants, he sent an order for $3,000 worth of stock to a nurseryman in France. He had no money for an irrigation system, and when there was a spell of dry weather, he nearly killed himself working day and night watering his plants by hand from a well which he had dug on the place. In the course of a few years, his bare, ugly hillside became a paradise of trees and shrubs, and he began distributing rare varieties of nut trees up and down the Pacific Coast." The business became so extensive and profitable that in 1882, he closed his barber shop and devoted his energies full time to the nursery, which later was named the Barren Hill Nursery. Gillet's knowledge of horticulture was recognized far and wide, and he was a regular contributor to newspapers and journals (11).

During the period 1880 to 1905, Gillet introduced nearly 20 cultivars of hazelnut. Most of these had little to recommend them. His 1880-81 catalogue lists four cultivars: 'Red Aveline,' 'White Aveline,' 'Grosse of Piedmont,' and 'Sicily.' The exact year of the introduction of the 'Barcelona' hazelnut is not known. It appears to be one of the early introductions, and 1885 is the date most often cited. It was first listed for sale under the name 'Barcelona' in the 1895 catalogue.

During that period of time, A.A. Quarnberg of Vancouver, Washington, was very active in the introduction and testing of hazelnuts in the Pacific Northwest. In a letter written by Quarnberg to Prof. C.I. Lewis, of the Oregon Agricultural College, in December, 1916, he states: "Mr. Gillet says that in the spring of 1886, Geo. A. Steel of Portland, Oregon, bought 165 second-generation seedlings together with 65 Red Aveline, 19 White Aveline, and 2 Purple Leaf

Aveline filbert trees. These probably were the first filbert trees brought to the Pacific Northwest.

"In a letter dated March 19, 1905, Mr. Gillet writes: 'The best for market I think are DuChilly Cobnut, Grosse Blanche of England, named by me Barcelona, etc., because, as he writes in other letters, it originated in Barcelona, Spain. Consequently, Mr. Gillet named the Barcelona, which fact is not generally known. I also find that I most probably planted the first DuChilly trees in the Northwest in 1894 and Henry Biddle of Clark County, the first Barcelona in 1895" (2).

Of the cultivars introduced by Gillet, 'Barcelona' had the greatest number of good characteristics (6) and was widely planted in the Willamette Valley. Other cultivars introduced by Gillet include 'DuChilly' and 'Montebello,' which are grown to a small extent, and 'Daviana' and 'Hall's Giant,' which are commonly used as pollinizers. George Dorris of Springfield, Oregon, another pioneer in the Northwest hazelnut industry, believed 'Daviana' to be the best pollinizer for 'Barcelona.' Schuster's pollination studies (13) verified that 'Daviana' was a superior pollinizer, and it has held that position for 50 years. Even today, the most commonly found older orchards in Oregon are 'Barcelona' with 'Daviana' pollinizers.

World importance and synonyms

The United States ranks fourth in hazelnut production behind Turkey, Italy, and Spain. Approximately 25,000 acres are currently bearing nuts (10). Ninety-eight percent of the U.S. production is in Oregon; the remaining comes from Washington. The majority of the hazelnut orchards are located in Oregon's Willamette Valley, where cool summers and mild winters prevail.

'Barcelona' is commercially important only in the United States and southwestern France, where it represents 82% and 51% of the trees, respectively. In Oregon, some growers prefer 'Ennis' over 'Barcelona' for the in-shell market, but 'Barcelona' continues to be widely planted. In France, 'Barcelona' represents only 10% of newer plantings. It has largely been replaced by 'Ennis' and to a lesser extent by 'Corabel.' Since the combined production of Oregon and France is roughly 3.2% of the world crop, 'Barcelona' accounts for only 2.4% of the hazelnuts grown in the world.

'Barcelona' is a very old cultivar and was probably widely distributed throughout western Europe in the nineteenth century. Gillet thought that it resembled nuts grown around Barcelona, Spain, and evidence exists to support this. Spain's production is centered around the town of Reus, along the Mediterranean Sea in the northeastern part of the country. However, 'Barcelona' is not commercially important there. Only a few trees can be found in orchards and home gardens throughout the region, where it is known as 'Castanyera' (12). In the northern province of Asturias, it is known as 'Grande' because of its large nut size.

In France, 'Barcelona' is known as 'Fertile de Coutard.' As in Spain, 'Barcelona' is widespread in backyards and gardens (3). According to Quarnberg's letter of 1916, Gillet imported the variety as 'Grosse Blanche d'Angleterre,' which translates as Large White of England. Spain is

In the course of a few years, his bare, ugly hillside became a paradise of trees and shrubs.

Said of Felix Gillet, whose work began the development of the modern hazelnut industry.

more likely the origin than England. 'Barcelona' is similar, if not identical to the cultivar 'Grosse Bunte Zellernuss,' a cultivar described and illustrated by Goeschke (1887) and for which he gave 'Barcelloner Zellernuss' as a synonym.

Description

'Barcelona,' like all of the world's major cultivars, has several desirable, as well as undesirable, characteristics. All things considered, it was the best of those introduced by Gillet. 'Barcelona' trees are moderately vigorous and precocious with a desirable growth of habit, being intermediate between upright and spreading. Like all European hazelnut trees, 'Barcelona' is monoecious and self-incompatible (14). Many S-alleles have been identified in the control of the sporophytic stigmas, but, because of dominance only the S1 allele is expressed in the pollen. Hazelnut trees bloom in midwinter and are wind-pollinated. Relative to other commercially available cultivars, 'Barcelona' sheds its pollen early and female flowers are receptive early.

Nuts are mostly borne in clusters of two or three. The husks are roughly 1.5 times as long as the nuts. At maturity, the nuts fall freely to the ground and can be harvested mechanically. The nuts are nearly round and average 3.2 grams when dried. 'Barcelona' outyields most of the older European cultivars, but is considered only moderately productive in Oregon. On the average, a mature 'Barcelona' orchard yields 0.8 tons/acre (.3 tonnes/hectare). 'Ennis' and 'Casina' will outyield 'Barcelona' by up to 50%. Alternate bearing is common.

'Barcelona' matures in mid-late season, and is generally harvested the first week of October. In Oregon, this late maturity is a disadvantage because it is often coincident with the start of the rainy season. Nuts which fall in mud are not only difficult to harvest, but soil particles adhere to the pubescence at the apical end of the nuts. As the soil is impossible to remove, the nut takes on a dull appearance.

Oregon's industry has traditionally produced large, nearly round, attractive nuts for the in-shell market. 'Barcelona' is well suited for this purpose. However, the in-shell market is quite limited, and currently half of the crop must be cracked and sold on the kernel market. 'Barcelona' is not suitable for the kernel market. The shell is moderately thick (43% kernel by weight) compared with the leading European cultivars, which average between 45 and 56%. Although the flavor is good, the raw kernels have a fibrous covering on the pellicle. Furthermore, since only half of the pellicle can be removed by dry heat (130°C for 14 min.), a process known as blanching, the product does not have the clean white appearance of the imported kernels of other cultivars. Also, the kernels are larger than desirable, and the texture is chewy rather than crisp. Another disadvantage to producers and processors is that 'Barcelona' yields a high percentage (25 to 40%) of defective nuts. The most common defect is blank nuts, i.e., shells containing no kernels. 'Barcelona' averages 10 to 25% blanks, depending on the year. Other defects include poorly developed, shriveled, and/or moldy kernels, and twins (two kernels in one shell). In certain years,

'Barcelona' is especially susceptible to a physiological disorder called "brown stain" which terminates kernel development and leads to blanks.

Diseases and pests

'Barcelona' is susceptible to bacterial blight caused by *Xanthomonas campestris* pv. *pruni,* a disease which is most serious in nurseries and young orchards. Although 'Barcelona' is moderately resistant to Eastern Filbert Blight, caused by the fungus *Anisogramma anomala,* trees will eventually succumb to it. This disease is native to the eastern United States, where the wild American hazel *(C. americana)* serves as its host. Introduced to southwestern Washington in the late 1960s, the disease has been spreading southward, and has now been reported throughout the northern end of the Willamette Valley.

'Barcelona' is highly resistant to the big bud mites, *Phytoptus avellanae* and *Cecidophyopsis vermiformis*. 'Daviana,' the principal pollinizer, is highly susceptible.

Breeding

Beginning early this century, in hopes of finding varieties superior to 'Barcelona,' several growers made selections such as 'Royal,' 'Ennis,' 'Butler,' 'Nonpareil,' 'Lansing #1,' and 'Jemtegaard #5' in populations of open-pollinated seedlings. Because 'Barcelona' and 'Daviana' have been the most widely planted cultivars, it is probable that these selections are hybrids of these two. Their morphological traits are consistent with those of seedlings from the controlled cross. Also, their incompatibility alleles support this assumption.

Each selection has one allele in common with 'Barcelona' and one in common with 'Daviana.' Of these selections, only 'Ennis' has been planted widely as a main-crop cultivar.

In the late 1950s, more than a hundred additional cultivars were imported from Europe and evaluated to find a cultivar better than 'Barcelona.' Because the vast majority of them were decidedly inferior, a hazelnut breeding program was initiated at Oregon State University in 1969. The main objective is to develop cultivars for the kernel market. Such cultivars would have smaller, rounder, thin-shelled nuts and round, crisp, easily blanched kernels. The nuts would mature earlier and have fewer defects than 'Barcelona.' Because of the recent spread of Eastern Filbert Blight into Oregon, resistance to this pathogen is now an important objective in the program. Big bud mite resistance is important not only because it can reduce yields, but also because blasted buds provide a port of entry for Eastern Filbert Blight.

Literature cited

1) Anonymous. 1908. Felix Gillet is called by death. *The Daily Morning News* of Grass Valley and Nevada City. January 28, 1908.

2) Anonymous. 1917. Early history of walnuts and filberts. *The Walnut Book and Horticultural Digest* 3(2):12-13.

3) Bergougnoux, F., E. Germain, and J.P. Sarraquigne. 1978. *Le noisetier—production et culture INVUFLEC,* Paris.

4) Bunyard, E.A. 1920. Cob-nuts and filberts. *J. Royal Hort. Coc.* 45:224-232.

5) Coote, G. 1898. The cultivation of the hazelnut. Oregon Agr. Exp. Sta. Bull. 52:1-7.

6) Crane, H.L., C.A. Reed, and M.N. Wood. 1937. Nut breeding. In: *USDA Yearbook of Agriculture.* U.S. Government Printing Office. pp. 827-889.

7) Fairchild, D.G. 1938. *The world was my garden; travels of a plant explorer.* C. Scribner's Sons, New York.

8) Goeschke, F. 1887. *Die Haselnuss—ihre Arten und ihre Kulture.* Verlag von Paul Parey, Berlin.

9) Lagerstedt, H.B. 1975. Filberts. *In:* J. Janick and J. N. Moore (eds.). *Advances in Fruit Breeding.* Purdue Univ. Press, West Lafayette, IN. pp. 456-489.

10. Moffett, M.H., K.K. Mohr, and S.J. Lawton. 1989. An industrial assessment of the Pacific Northwest hazelnut industry. Report to the Small Business Administration. pp. 1-49.

11. Parsons, C.E. 1962. Felix Gillet. Nevada County Historical Society 16(4):1-3.

12) Rovira, M. 1988. Descripicio de varietats d'avellaner. Servicio Agropecuario Provincial de Tarragona, Reus.

13) Schuster, C.E. 1924. Filberts: 2. Experimental data on filbert pollination. Oregon Agr. Exp. Sta. Bull. 208.

14) Thompson, M.M. 1979. Incompatibility alleles in *Corylus arellana* L. cultivars. *Theor. Appl. Genet.* 55:29-33.

Authors Mehlenbacher and Miller were with the Department of Horticulture, Oregon State University, Corvallis, Oregon, when this article originally appeared, January 1986.

The 'Elberta' Peach

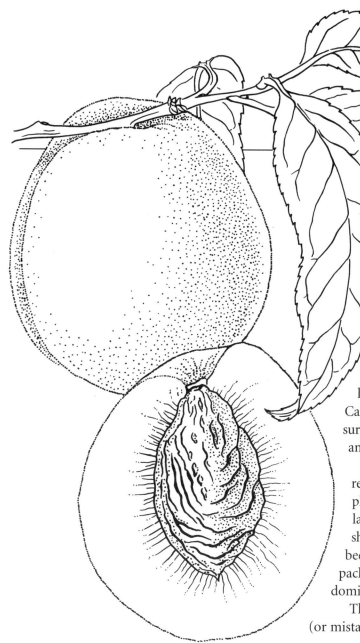

by Stephen C. Myers, W.R. Okie, & Gary Lightner

It is noteworthy that 1989 marks the one-hundredth anniversary of the American Pomological Society placing 'Elberta' peach on its list of recommended fruit (9). For much of the twentieth century, 'Elberta' dominated the commercial peach industry in the United States (1, 5, 7, 9, 16). In the years from 1910-1930, when the Georgia peach industry peaked at 16 million trees, about 40% of the production was of 'Elberta.' At that time, 'Elberta' was the only yellow-fleshed peach in the top eight cultivars (11). It started the shift from white to yellow-fleshed peaches. As late as 1965, 'Elberta' was still in the top ten peach cultivars in Georgia. In 1950, 45% of South Carolina's 4.5 million peach trees were 'Elberta,' down from 60% ten years earlier (18). A 1968 survey showed 'Elberta' in fifth place in Maryland (down from first in 1956), first in Pennsylvania, and first in Virginia (12).

Most people associated with stone fruit culture would agree that 'Elberta' has been replaced and surpassed by superior cultivars. However, few would question that 'Elberta' played a pivotal role in the development of contemporary peach and nectarine culture. In large part, the cultivar's shipping characteristics signaled the beginning of the modern peach shipping industry, creating new production areas in the early 1900s which, theretofore, had not been close enough to major markets to prosper. Concurrent advances in transportation, packaging, and cooling complemented the cultivar (16), which moved from obscurity (8) to dominance (1, 5, 7, 9, 16) in a relatively short period of time.

The dominance of 'Elberta' also formed a lasting imprint in the public's mind. 'Elberta' (or mistakenly, 'Alberta') continues to have strong name recognition at the garden center and at the fruit stand. Few, save 'Georgia Belle' or 'Redhaven,' have such name recognition, particularly interesting in a fruit which, unlike apple, has few visible characteristics to distinguish cultivars.

Research with 'Elberta' has also left a lasting legacy on the body of knowledge which makes up our current understanding of peach growth and development. Scientific contributions utilizing 'Elberta' established important standards and principles still in use today (6, 22). Some of the more common include dormancy and rest requirements, critical temperatures for bud hardiness, influence of numerous cultural practices on fruit yield and quality, as well as fundamental principles of fruit maturity. In many areas, maturity dates for cultivars are still commonly described in terms of maturity a certain number of days before or after 'Elberta' (6).

The history of 'Elberta' was recorded in detail, due in part to documentation by the Georgia Horticultural Society, an organization of significant size and prominence around the time 'Elberta' was selected (9, 16, 19). Dr. P.J.A. Berckmans, a noted pomologist (he collaborated with Charles Downing in preparing second and third editions of *The Fruit and Fruit Trees of America)*, innovator in peach production and shipping, and founder of the Georgia Horticultural Society, was himself president of the American Pomological Society from 1887 to 1897.

The origin of 'Elberta' is fascinating and serendipitous (9, 16). Robert Fortune, an English botanist who had been sent to China by the London Horticultural Society to collect plants, sent seeds and a potted tree of a delicious peach growing south of Shanghai to England in 1844 under the name of 'Shanghai.' This peach was probably the old cultivar now known in China as 'Shanghai Shuimi' (23). 'Chinese Cling' was imported in 1850 to the United States as potted trees labeled 'Chinese Cling' or 'Shanghai' by Charles Downing through a Mr. Winchester, British consul in Shanghai. The trees of the two cultivars were apparently identical. Downing sent one of the trees to Henry Lyons of Columbia, South Carolina, with whom the cultivar first fruited in the United States in 1851.

During the mid-1850s, records show that a Mr. L.C. Plant, a progressive banker in Macon, Georgia, had a secondary interest in fruit growing. In 1857, a Delaware nursery salesman stopped by Mr. Plant's Macon bank and convinced him to try some budded peach trees. Prior to that time, most people in Georgia had been producing their trees from seed even though budded trees were available. Mr. Plant placed an order for a few trees of 'Chinese Cling,' 'Early Crawford,' 'Late Crawford,' 'Oldmixon Free,' and 'Stump-the-World.' Mr. Plant sent these budded trees to his good friend Colonel Lewis Rumph of Marshallville, Georgia, a small town 35 miles southwest of Macon. Colonel Rumph grew these trees in the family orchard, and, with time, decided that fruit from 'Chinese Cling' were especially good. Being in such a family orchard, blooms of 'Chinese Cling' were subject to open pollination by other cultivars in the planting. Colonel Rumph's wife saved seeds from the 'Chinese Cling' tree and gave them to her grandson, Samuel H. Rumph. He planted the seeds out on the Rumph farm in

1870. Of the seedlings which developed in this planting, a number produced excellent fruit, one of which was Mr. Rumph's favorite and eventually came to be named 'Elberta.'

Samuel H. Rumph married Miss Clara Elberta Moore, a charming lady who entertained numerous friends. During one of Mrs. S.H. Rumph's "spend-the-day" parties, Samuel was showing the guests some of his choice peaches from seedlings along with others and announced each by cultivar name. He at last showed what he considered to be the best peach of all but gave no name. One of the guests, Mrs. L.E. Veal, inquired of the name. Mr. Rumph replied, "It is new, it has no name. You may name it." With that, Mrs. Veal replied, "Well, let's honor your wife and call it for her. She is perfect, and so is the peach. You will never have anything on this continent to surpass it. 'Elberta' is its name. Thanks for the honor."

At the time, Mr. Rumph speculated that the 'Chinese Cling' bloom that produced 'Elberta' had been fertilized by 'Early Crawford.' However, out of 2,200 open-pollinated and selfed seedlings of 'Elberta,' Palmer (13) found that none resembled 'Early Crawford.' He suggested that 'Elberta' was a natural selfed seedling of 'Chinese Cling,' with recessive yellow flesh breeding true for that color (7, 13).

Palmer's theory is unlikely in light of current knowledge of peach genetics. Although it is not possible to verify the characteristics of the 'Chinese Cling' tree grown by Colonel Rumph, later descriptions call it a white-fleshed clingstone with reniform leaf glands and showy, pollen-sterile flower. A genetic clingstone cannot produce a freestone seedling without cross-pollination. On the rare occasions when this sterile peach produced a self-pollinated fruit, the seedling would have showy, sterile blooms, in contrast to those of 'Elberta' or 'Georgia Belle.'

As Table 1 shows, 'Chinese Cling' could have crossed with 'Early Crawford' to produce 'Elberta' and with 'Oldmixon Free' to produce 'Georgia Belle,' if it was heterozygous for the gene for yellow flesh. However, 'Late Crawford' and 'Stump-the-World' would be listed just like 'Early Crawford' and 'Oldmixon Free,' respectively, and could be parents of either 'Elberta' or 'Georgia Belle.' The presence of the gene for pollen sterility carried by both 'Elberta' and 'Georgia Belle' reinforces their claim to being descendants of 'Chinese Cling,' since pollen sterility was undescribed before being noticed in seedlings of 'Georgia Belle' and later 'Elberta.' 'Chinese Cling' is probably the oldest American peach known to be pollen sterile.

To test the claim that 'Chinese Cling' sired both 'Elberta' and 'Georgia Belle,' a small progeny of 'Chinese Cling' (this clone matches early published descriptions of the cultivar) was fruited at Byron in 1988. The seedlings included 11 white-fleshed clingstones, ten white-fleshed freestones, eight yellow-fleshed clingstones, and seven yellow-fleshed freestones. 'Chinese Cling' must be carrying the recessive gene for yellow flesh. Otherwise, all offspring would have been white-fleshed. The clingstone seedlings probably resulted from outcrosses to adjacent 'Babygold 5' trees, which are clingstone.

Mr. Rumph felt that the new cultivar would

> *"Well, let's honor your wife and call it for her. She is perfect, and so is the peach. You will never have anything on this continent to surpass it. 'Elberta' is its name."*
>
> Mrs. L.E. Veal to Samuel H. Rumph on the naming of 'Elberta.'

TABLE 1.

Characteristics of 'Elberta' peach and its supposed relatives, with possible genotypes in parentheses.

Cultivar	Flesh color	Pit adherence	Flower type	Leaf gland	Pollen viability
Early Crawford	yellow (yy)	free (F_)	non-showy (ShSh)	globose (Ee)	fertile (PsPs)
Elberta	yellow (yy)	free (Ff)	non-showy (Shsh)	reniform (EE)	fertile (Psps)
Chinese Cling	white (Yy)	cling (ff)	showy (shsh)	reniform (EE)	sterile (psps)
Georgia Belle	white (Yy)	free (Ff)	non-showy (Shsh)	reniform (EE)	fertile (Psps)
Oldmixon Free	white (Y_)	free (F_)	non-showy (ShSh)	globose (Ee)	fertile (PsPs)

TABLE 2.

Cultivars containing 'Elberta' in their genetic background, assuming 'J.H. Hale' as the progeny of unknown parents.

Adria	Fireprince	La Jewel	Springold
Amador[2]	Firered	La Premiere	Stark Compact[2]
Amrein[2]	Flamecrest	La Red	Stark Lateglo[2]
Anza	Flavorcrest	Loring	Stark Late Gold
Arp Beauty[2]	Frank[2]	Margaret Kane	Stark Saturn
Babdon	Frankie	Marglow	Starlite
Bicentennial	Fulmur[2]	Marigold	Sullivan Elberta[2]
Biscoe	Garden State[2]	Mark-Berta	Summergold
Blake	Gloribloom	Mark-Late	Summer Pearl
Bonette[2]	Goldcrest	May Crest	Sunbeam
Bonita	Golden Beauty	Maydon	Sungold
Bounty	Goldeneast[2]	Maywel	Sunprince
Brayberta[2]	Golden Flame	McNeely	Sunrich
Buttercup	Golden Globe	Missouri	Surecrop
Calred	Golden Jubilee	Norman	TAMU Denman
Camden	Golden State	Quachita Gold	Telford[2]
Canadian Harmony	Goldray	Ozark	Topaz
Canadian Queen[2]	Gurney's Dakota[2]	Pacemaker	Triogem
Carolina Belle	Harbrite	Poppy	Troy
Carrie	Harcrest	Prairie Dawn	Tulip
Casella Queen	Harken	Prairie Rambler[2]	Valiant[4]
Chadon	Harland	Prairie Schooner[2]	Vanderpoole[2]
Chaffey[3]	Harrow Beauty	Prairie Sunrise	Vanguard
Christensen Early Elberta[2]	Harrow Diamond	Primrose[3]	Vanity
Comanche	Harson	Redelberta[2]	Vedette[2]
Cullinan	Harvester	Redglobe	Vedoka
Derby	Hickman's Elberta[2]	Redqueen	Veefreeze
Donwel	Honeyberta[3]	Redskin[3]	Velvet
Early Fair Beauty	Howard Fisher	Roberta[2]	Vesper
Early Triogem	Jefferson	Romance	Vimy[2]
Elberta Queen[2]	Jerseyglo	Royal[3]	Welberta[3]
Emery	Jerseyqueen	Ruston Red	Weldon
Envoy	Jubilant	Salberta[3]	Wilma[2]
Erlyvee	Jun-Berta	Scott Elberta[2]	Winblo
Fair Beauty[2]	Ken Late Elberta[2]	Sentry	Yelo
Fayette	Kette[2]	Sessen Cling[2]	
Fireglow	La Gem	Springcrest	

[2] Elberta—female parent
[3] Elberta—male parent
[4] Elberta x Elberta cross

withstand shipping, previously a limiting factor in commercial production. In a trial shipment of 'Elberta,' packed in one-third-bushel crates, fruit arrived at a distant market in good condition with no refrigeration. These peaches brought five dollars per crate or fifteen dollars per bushel. The first major commercial shipment of peaches out of Georgia were grown by Mr. Rumph at his Willow Lake Orchard and Nursery. He is also credited with development in 1875 of a peach shipping refrigerator and of the rigid mortised-end peach crate. Considered father of the Georgia commercial peach industry, his accomplishments are today noted by a historical marker at his home in Marshallville.

In 1870, Mr. Lewis A. Rumph, son of Colonel Lewis Rumph, planted some seeds from the same 'Chinese Cling' tree that produced 'Elberta.' From those seedlings, he selected and named 'Belle,' listed by the American Pomological Society in 1899 as 'Georgia' but changed to 'Belle' in 1909. Popularly, it came to be called 'Georgia Belle.' L.A. Rumph speculated that it was a cross of 'Chinese Cling' and 'Oldmixon Free.' The sites of the original 'Elberta' and 'Belle' trees are marked at the Rumph farm by a Georgia historical marker.

At the time of its introduction, many attributes were listed for 'Elberta' (9). Its adaptability to a broad range of soil and climatic conditions resulted in its being "grown in every peach-growing state in the Union…" (9). Trees were cited to be long-lived and known for consistent annual production. Trees, described as large vigorous, produce an upright spreading, dense-topped crown. Leaves are dark olive-green, margin fine to coarsely serrate with one to six reniform glands. Rest requirement for flower buds is 850 hours and for leaf buds 950 hours (21).

Fruit of 'Elberta' were described in 1917 as "large, handsome, well-flavored fruits which ship and keep remarkably well" (9). They had a thick skin and ripened more slowly than older cultivars (4). Fruit, which mature in midseason with one-fourth to three-fourths surface red overcolor, are yellow-fleshed freestone with a sweet or subacid taste.

However, 'Elberta' has serious faults which may have limited its use had it not been such an excellent shipping peach, a quality superior to all others available at the time for the commercial trade (9, 16, 20). Even in early descriptions, 'Elberta' was described to "fall short in quality" (9). Fruit have a pronounced bitterness or astringency even when peaches are fully ripe. The astringency is particularly strong in cooler climates (20). Hedrick (9) wrote, "Picked green and allowed to ripen in the markets, 'Elberta' is scarcely edible by those who know good peaches." By today's standards, 'Elberta' has an unattractive exterior, drops badly as it approaches maturity, and is not resistant to flesh browning (15). In addition, the stone is large. Irrespective of these shortcomings, the positive attributes of 'Elberta,' particularly shipping characteristics, were great enough to ensure its utility as a commercial peach for some time (9).

As a parent, 'Elberta' transmitted large fruit size, thick skin, firmness, yellow flesh freestone character, and a prolonged ripening period to offspring (2, 7). However, a shortcoming of 'Elberta' is that it is lacking in wood and bloom hardiness. Hedrick (9) noted that its "blossoms open rather too early in New York." The noted fruit breeder M.A. Blake (1883-1947) at New Jersey was aware of this characteristic in 'Elberta' and objected to 'Elberta' as a parent because it transmitted lack of hardiness (2, 7, 19). However, he did develop varieties with considerable hardiness by crossing 'Elberta' with more hardy varieties. J.H. Weinberger, a most successful peach, nectarine, and grape breeder in Georgia and California, noted that, as a parent, 'Elberta' has "turned out very few good varieties" (20). Early on, he accepted the experience of breeders before him that 'Elberta' was not a good parent for breeding programs. Self-pollinated seedlings were found to show better quality than 'Elberta' itself (7). The late Stanley Johnston (1893-1963), noted breeder of the Haven series at the South Haven, Michigan, Experiment Station, found that 'J.H. Hale' was a much better parent than 'Elberta.' However, it is noted that 'Elberta' is likely one of 'J.H. Hale's' parents (5, 9, 10, 16, 17). In part, this conclusion is made because of large numbers of similarities between 'Elberta' and 'J.H. Hale' (7). Also, pollen sterility was unknown until noticed in progenies of 'Georgia Belle' and later of 'Elberta' (7).

'J.H. Hale' was discovered as a single tree in a lot of 'Early Rivers' peaches shipped by David Baird of Manalapan, New Jersey, to J.H. Hale and planted on his farm at South Glastonbury, Connecticut. Trees propagated from this one performed well on Hale's farm in Fort Valley, Georgia. In 1912, Hale sold the rights to W.P. Stark

> *'Elberta' has serious faults which may have limited its use had it not been such an excellent shipping peach.*

TABLE 3.

Cultivars containing 'Elberta' in their genetic background, assuming 'Elberta' as parent of 'J.H. Hale,' and number of occurrences in their ancestry.

Cultivar	#	Cultivar	#	Cultivar	#
Adria	5	Early Coronet	2	Golden Glory	1
Afterglow	1	Early East	1	Golden Jubilee	1
Albru	1	Early Fair Beauty	1	Golden Monarch	2
Allgold	2	Early Hale Haven	1	Golden State	1
Amador	1	Early Raven	1	Golden Supreme	3
Amrein	1	Early Re&aven	2	Goldenest	2
Angelus	2	Early Triogem	2	Goldenred	1
Anza	2	Earlytop	1	Golderest	3
Arp Beauty	1	Eden	1	Goldgem	1
Aurora	2	Elberta Queen	1	Goldilocks	2
Autumn	1	Ellerbe	5	Goldray	1
Babdon	2	Emery	1	Goodcheer	2
Babygold 5	1	Empress	2	Gurney's Dakota	1
Babygold 6	2	Envoy	2	Gypsy	1
Babygold 7	1	Erlyvee	1	Halehaven	1
Bicentennial	4	Eve	2	Hamlet	5
Biscoe	3	Fair Beauty	1	Harbelle	8
Blake	2	Fairhaven	1	Harbinger	2
Bonette	1	Fairlane	2	Harbrite	6
Bonita	2	Fairway	2	Harcrest	4
Bonjour	2	Fallate	1	Harken	6
Brandywine	2	Fantasia Nectarine	2	Harland	6
Brayberta	1	Fayette	4	Harmony	2
Brighton	3	Fertile Hale	1	Harrison	1
Buttercup	1	Fillette	1	Harrow Beauty	11
Calred	2	Fireglow	2	Harrow Diamond	3
Camden	4	Fireprince	7	Harson	6
Canadian Harmony	6	Firered	4	Harvester	2
Canadian Queen	1	Flamecrest	7	Havis	4
Candor	2	Flavorcrest	4	Hermosa	1
Cardinal	2	Flordabeauty	5	Hermosillo	1
Carrie	1	Flordabelle	2	Hickman's Elberta	1
Casella Queen	2	Flordagold	2	Hiland	3
Catherina	5	Flordaqueen	2	Home Canner	1
Chadon	3	Flordared	2	Honey Dew Hale	1
ChaXey	1	Flordawon	2	Honeyberta	1
Cherryred	1	Fortyniner	1	Honeygem	1
Christensen Early Elberta	1	Franciscan	1	Howard Fisher	2
Clark	1	Frank	1	Improved Pacifica	1
Clayton	5	Frankie	1	J.H. Hale	1
Collins	3	Frostqueen	1	Jefferson	3
Columbina Nectarine	2	Fujihara Babcock	1	Jerseydawn	2
Comanche	2	Fulmur	1	Jerseyglo	5
Compact Redhaven	2	Gaiety	1	Jerseyland	1
Constitution	1	Garden State Nectarine	1	Jerseyqueen	1
Coronet	2	Garnet	1	Jubilant	1
Correll	5	Garnet Beauty	2	July Lady	2
Cresthaven	5	Gemfree	1	Jun-Berta	1
Cullinan	4	Glohaven	2	June Bride	1
Derby	5	Gloribloom	2	June Lady	2
Desertgold	10	Gold King Nectarine	1	Juneprince	7
Dixiland	1	Gold Rush	1	Kalhaven	1
Dixired	2	Golden Babcock	2	Ken Late Elberta	1
Donwel	2	Golden Beauty	1	Kette	1
Earlired	4	Golden Flame	2	Keystone	3
Early Amber	2	Golden Globe	2	Kim Earling	1

Nursery, which rapidly commercialized it (7). The popularity of 'Elberta' for canning resulted in a large quantity of seed being available, hence it was often used as a rootstock for budding. It is possible that 'J.H. Hale' was an unbudded 'Elberta' seedling that was not rogued from the nursery row and was sold as a budded tree.

Interestingly, the senior author attained one source of this reported relation between 'Elberta' and 'J.H. Hale' from Dr. J.H. Weinberger (20), who in turn was told the story by a Mr. John H. Baird, owner of Georgia peach land leased to the United States Department of Agriculture in the 1930s. Previously, Mr. Baird worked for and bought the land from Mr. J.H. Hale. Mr. Hale had tested the 'J.H. Hale' peach on that land in earlier times and conveyed directly to Mr. Baird his opinion that 'J.H. Hale' peach was a seedling of 'Elberta.' The connection between 'Elberta' and 'J.H. Hale' will remain speculative, an academic point that may one day be answered with the use of genetic mapping (14). In any event, over time, both 'Elberta,' and especially, 'J.H. Hale' have been useful to breeders (5, 7, 17) in creating better varieties, and 'Elberta' is found in a large percentage of pedigrees of common cultivars (Table 2).

The coefficients of coancestry of 'Elberta' crosses are generally high due to the presence of 'Elberta' in the ancestry of many cultivars (17). For example, first cousins have a coefficient of 0.063. The average coefficient of 'Elberta' crossed with 30 popular cultivars averaged 0.059. Both 'Elberta' and 'J.H. Hale' transmit a lack of cold hardiness to their progeny (2, 3, 7), probably because they both descend from the southern group of Chinese peaches (17). The majority of modern-day peaches descend from a small group of ancestors, in fact a very narrow gene pool (17). If 'Elberta' is indeed 'J.H. Hale's' parent, it would influence the former's role in the ancestry of peach cultivars (Table 3). The average coefficient rises to 0.218 if 'J.H. Hale' is considered an 'Elberta' offspring (17).

It is fascinating today that a cultivar with so few noteworthy characteristics could have shaped an industry and left such an impact. However, in its day, 'Elberta' was simply, as Dr. Weinberger (20) noted, "way ahead of it's time." Most peaches available at the time, many from Europe, were developed for local consumption, possessed high quality white flesh, but softened quickly and were not suitable for shipping. There was "nothing to compare with it at the time" (20). The times were right for the development of a shipping industry, with recent advances in handling, cooling and transportation. 'Elberta' filled the niche as no other variety at the time could. The peach season in a given peach production area became a three-week "'Elberta' season" (20). The "'Elberta' season" moved up through the country from south to north. Brokers had no difficulty because they knew what to expect—more 'Elberta.' With the introduction of earlier varieties, this pattern of production and marketing began to change (1, 7, 10). Likewise, introduction of superior quality cultivars eventually spelled the demise of 'Elberta's' supremacy and importance as a commercial peach. However, 'Elberta's' contributions leave little doubt that its place in the history of peach culture is well secured.

TABLE 3, continued.

Cultivar	Count
Kim Nectarine	1
Kirkman Gem	1
La Gem	4
La Gold	2
La Premiere	3
La Red	2
Late Le Grand Nectarine	1
Late Sunhaven	4
Laterose	1
Le Grand Nectarine	1
Loring	2
Madison	2
Magnolia	2
Mardigras	2
Margaret Kane	1
Marglow	2
Marhigh	1
Marigold	1
Mark-Berta	1
Mark-Late	2
Marland	1
Marpride	1
Marqueen	1
Marsun	1
May Crest	3
May Lady	1
Maybelle	2
Maydon	1
Maygold	3
Maytime	1
Maywel	2
McNeely	3
McRed	3
Merrill Gem	1
Merrill Hale	1
Merrill Prince	1
Merritt	1
Midway	1
Milam	4
Missouri	1
Monroe	1
Mountaingold	1
Necta-Heath	1
Newcheer	2
Newday	1
Norman	3
Opedepe	3
Ouachita Gold	4
Ozark	2
Pacemaker	2
Pacifica	1
Pat's Redhaven	2
Pekin	3
Piedmontgold	1
Poppy	2
Prairie Clipper	1
Prairie Dawn	3
Prairie Daybreak	1
Prairie Rambler	1
Prairie Rose	1
Prairie Schooner	1
Prairie Sunrise	3
Prenda	1
Primrose	1
Ramsey	1
Ranger	2
Raritan Rose	1
Rayon	2
Red Elherta	1
Red Gold	1
Red Grand Nectarine	3
Red King Nectarine	1
Red Lady	1
Redcap	4
Redglobe	2
Redhaven	2
Redqueen	3
Redrose	1
Redskin	2
Redtop	1
Regina	1
Richhaven	4
Rio Grande	2
Roberta	1
Romance	1
Rosydawn	2
Royal	2
Royal Gem	1
Roza	1
Rubired	4
Ruston Red	5
Salberta	1
Schooldays	1
Scott Elberta	1
Sentinel	2
Sentry	4
Sessen Cling	1
Shepard's Beauty	2
Shermans Red	3
Shoji	2
Sixty-six	1
Solo	1
Somervee	1
Southland	2
Sparkle	1
Splendor	1
Springbrite	3
Springerest	4
Springold	4
Stark Earliglo	1
Stark Encore	1
Stark Late Gold	1
Stark Lateglo	1
Stark Saturn	3
Starkoompact	1
Starlite	4
Sullivan Early Elberta	1
Summer Pearl	2
Summercrest	1
Summerglo	3
Summergold	2
Summerqueen	1
Summerrose	1
Summerset	1
Sun Grand Nectarine	1
Sun Lady	1
Sunbeam	1
Sunbrite	8
Suncling	1
Sunfre	2
Sungold	1
Sunhaven	4
Sunhigh	1
Sunlite Nectarine	2
Sunnyside	10
Sunprince	3
Sunrich	1
Sunripe	1
Sunrise	1
Sunshine	2
Superior	1
Surabian	1
Surecrop	4
Suwanee	2
Tamu	2
Telford	1
Thomason Early Elberta	2
Topaz	4
Triogem	2
Tropic Sweet	1
Troy	3
Tulip	1
Tyler	1
Valiant	2
Vanderpoole	1
Vanguard	2
Vanity	3
Vedette	1
Vedoka	2
Veefreeze	2
Veeglo	1
Velvet	4
Vesper	2
Vimy	2
Washington	1
Welberta	3
Welcome Hale	1
Weldon	2
White Hale	1
Wildrose	1
Wilma	1
Winblo	4
Yellow King	1
Yelo	3
Zachary Taylor	1

Literature cited

1) Auchter, E.C. and H.B. Knapp. 1937. *Orchard and Small Fruit Culture.* John Wiley and Sons, Inc., New York.

2) Blake, M.A.1933. 'Elberta' and its selfed and chance seedlings lack hardiness. New Jersey Agr. Exp. Sta. Circ. 287.

3) Blake, M.A. 1934. Additional facts in regard to the 'J.H. Hale' peach as a parent in breeding work. *Proc. Amer. Soc. Hort. Sci.* 30:124-128.

4) Blake, M.A. and L.J. Edgerton. 1946. Breeding and improvements of peach varieties in New Jersey. Rutgers University Bulletin 726.

5) Chandler, W.H. 1942. *Deciduous Orchards.* Lea and Febiger, Philadelphia, Pennsylvania.

6) Childers, N.F. and W.B. Sherman (eds.). 1988. *The Peach.* Horticultural Publications, New Jersey.

7) Cullinan, F.F. 1937. Improvement of stone fruits, pp. 665-702. In: *USDA Yearbook of Agriculture.*

8) Fulton, J.A. 1892. *Peach Culture.* Orange Judd Company, New York.

9) Hedrick, U.P. 1917. *The Peaches of New York.* J.B. Lyon Co., Printers, Albany, New York.

10) Iezzoni, A.F. 1987. The 'Redhaven' peach. *Fruit Varieties Journal* 41(2):50-52.

11) McHatton, T.H. and H.W. Harvey. 1919. Peach growing in Georgia. Georgia State College of Agriculture Bulletin 169.

12) Oberle, G.D. 1968. Variety revolution in the East. *Proc. Nat. Peach Convention* 27:31-37.

13) Palmer, E.F. 1923. Solving the Fruit Grower's Problems by Plant Breeding. *Amer. Soc. Hort. Sci. Proc.* 19:115-124.

14) Rom, R.C. 1989. Personal Communication. Department of Horticulture and Forestry, Univ. of Arkansas, Fayetteville, Arkansas.

15) Savage, E.F. and V.E. Prince. 1972. Performance of Peach Cultivars in Georgia. University of Georgia Agri. Exp. Sta. Bull. 114.

16) Savage, E.F. 1989. Personal Communication. Box 435, Experiment, GA 30312.

17) Scorza, R., S.A. Mehlenbacher, and G.W. Lightner. 1985. Inbreeding and Ancestry of Freestone Peach Cultivars of the Eastern United States and Implications for Peach Germplasm Improvement. *J. Amer. Soc. Hort. Sci.* 110(4):547-552.

18) Senn, T.L. and J.S. Taylor. 1951. The commercial peach industry in South Carolina. Clemson Agricultural College Bull. 393.

19) Stuckey, H.P. 1945. Diamond Jubilee of the Elberta Peach. *American Fruit Grower* 1:8, 30, 44.

20) Weinberger, J.H. 1989. Personal Communication. 6111 East Lyell, Fresno, CA.

21) Weinberger, J.H. 1950. Chilling requirements of peach varieties. *Proc. Amer. Soc. Hort. Sci.* 56:122-128.

22) Westwood, M.N. 1978. *Temperate-zone Pomology.* W.H. Freeman and Company, New York.

23) Zai-long, Li. 1984. Peach germplasm and breeding in China. *HortScience* 19:348-351.

Authors Myers, Okie, and Lightner were with the Department of Horticulture, University of Georgia, Athens; USDA, Southeastern Fruit and Tree Nut Research Station, Byron, Georgia; and USDA-ARS, Appalachian Fruit Research Station, Kearneysville, West Virginia, respectively, when this article originally appeared, October 1989.

The 'Bluecrop' Highbush Blueberry

by Arlen Draper & Jim Hancock

Domesticated highbush blueberry culture began in the early 1900s in New Jersey through the efforts of Dr. F.V. Coville of the U.S. Department of Agriculture, and Miss Elisabeth C. White. The first artifically hybridized cultivar was released by Dr. Coville in 1920. The highbush acreage has now grown to over 17,000 hectares, with most being located in Michigan (7,500 ha), New Jersey (3,700 ha), British Columbia (2,000 ha), and North Carolina (1,500 ha) (5).

Over 70 cultivars have been released by public breeders since 1920. Of these, the 'Bluecrop' is by far the most widely grown cultivated blueberry cultivar in the world. It is grown in all highbush producing areas of the United States and Canada, and is being planted rapidly in all other countries that are developing commercial acreages. The United States is the world's largest producer of this native American fruit, and 'Bluecrop' is the most important cultivar (5).

'Bluecrop,' tested as 17-19, originated from a cross of GM-37 ('Jersey' x 'Pioneer') x CU5 ('Stanley' x 'June') made in 1934 by F.V. Coville, United States Department of Agriculture, and O.M. Freeman, New Jersey Agricultural Experiment Station. In its genetic background are the wild selections 'Brooks,' 'Grover,' 'Sooy,' 'Rubel,' and 'Russel' (6).

The original seedling of 'Bluecrop' was grown and selected at Weymouth, New Jersey in 1941 by George M. Darrow, USDA, and J.H. Clark, New Jersey, Agricultural Experiment Station. A joint release (USDA-NJ) of 'Bluecrop' to commercial growers was made in December, 1952 (1). 'Bluecrop' had been tested quite extensively prior to its introduction; some growers and researchers were not impressed by its sparse foliage, which appears incapable of maturing the heavy crops, and rejected it. That proved to be a costly decision.

The original description (1) of 'Bluecrop' when introduced proved to be accurate: "Fruit cluster large and medium loose,

A History of Fruit Varieties

> *'Bluecrop,' though an outstanding cultivar, has shown only modest potential as a parent in breeding pure highbush blueberries.*

Authors Draper and Hancock were with the USDA Fruit Laboratory, Beltsville, Maryland; and the Department of Horticulture, Michigan State University, East Lansing, Michigan, respectively, when this article originally appeared, January 1990.

berries roundish-oblate, color very light blue, very firm, subacid, flavor good, moderately aromatic, scar small, ripens in mid-season, stem sometimes clings to berry. Bush upright and vigorous, leaves medium to below medium in size, very consistent producer." Though 'Bluecrop's' productivity in individual years is not always greater than other cultivars, its outstanding feature is consistent yearly production with a mean annual yield of about three kilograms per plant over a ten-year period in Michigan and Arkansas (3). Its only really negative factors are its tendency to produce tart fruit and lose its upright habit under high fruit loads.

In addition to its genetic potential for yields, some contributing factors to 'Bluecrop's' success are cold hardiness, drought tolerance, and disease resistance (4). One of its most outstanding features is broad soil and climatic adaptation, enabling it to be grown from New Jersey, north and west to Michigan, south to Arkansas, east through north Mississippi and Tennessee to North Carolina, north to New Jersey, and in the Pacific Northwest, including British Columbia, Canada. It is also grown in several European countries.

'Bluecrop,' though an outstanding cultivar, has shown only modest potential as a parent in breeding pure highbush blueberries. The 'Darrow' highbush blueberry, grown on a limited basis because of low winter hardiness, is the only North American highbush cultivar with 'Bluecrop' as a parent (2). 'Bluecrop' is also a parent of the New Zealand cultivars 'Puru,' 'Nui,' and 'Reka,' which came from the cross of E-118 x 'Bluecrop.' Selection E-118 originated from a cross of 'Ashworth' (wild *V. corymbosum*) x 'Earliblue.'

'Bluecrop' has been extremely useful in germplasm enhancement efforts using interspecific hybrids. It crosses readily with unreduced gamates of many diploid species (particularly *V. darrowi*), and these complex hybrids have been important in southern highbush blueberry breeding. It is a grandparent of 'Cape Fear,' 'Blue Ridge,' 'Cooper,' and 'Gulfcoast,' and is both a maternal and paternal grandparent of 'Georgiagem.'

Though 'Bluecrop' is not the perfect blueberry cultivar, it, more than any other because of good fruit quality and consistent bearing, enabled growers to meet annual consumer demands and turn blueberry growing into a legitimate commercial enterprise. It is still highly recommended for planting in all highbush growing areas. Blueberry plantings remain productive for many years, and this will ensure that 'Bluecrop' will continue to dominate the early midseason cultivar scene well into the next century.

Literature cited

1) Brooks, R.P. and H.P. Olmo. 1953. Fruit and nut register, list number 8. *Proc. Amer. Soc. Hort. Sci.* 62:517.

2) Brooks, R.P. and H.P. Olmo. 1972. Register of new fruit and nut varieties. 2nd edition. University of California Press, Berkeley.

3) Eck, P. 1988. Blueberry science. Rutgers Univ. Press, New Brunswick.

4) Eck, P. and N.F. Childers. Blueberry culture. Rutgers Univ. Press, New Brunswick.

5) Hancock, J. and A. Draper. 1988. Blueberry culture in North America. *HortScience* (in press).

6) Hancock, J. and J. Siefker. 1982. Levels of inbreeding in highbush blueberry cultivars. *HortScience* 17:363-366.

The 'McIntosh' Apple

by John T.A. Proctor

Ask a Canadian to name one apple cultivar and the response will be 'McIntosh.' A two-hundred-year history in Canada, the eastern United States, and parts of Europe, particularly Scotland, has ensured ready recognition and acceptance of this cultivar in spite of challengers (11). 'McIntosh' originated as a chance seedling on the John McIntosh homestead (see Figs. 1-3) near Dundela, Dundas County, Ontario, where John's son, Allan, began its propagation in the nursery about 1870 (1). Since that time, it has been grown widely and is most successful in climatic zones similar to its place of origin. Clear days and cool nights prior to harvest are prerequisites for good color and quality, and successful storage.

The original 'McIntosh' was a blush-type apple (14). Streak-type mutants showed up in the 1930s. Since then, every attempt has been made by nurserymen and researchers to ensure that only the more attractive blush types are offered to consumers. The search continues. In the 1960s, Fisher (2) discovered some spur-type mutants of 'McIntosh' in British Columbia and this sparked new interest in these and other strains (Table 1).

The spur-type growth habit was the major reason for the selection of mutants in British Columbia (2). Some of these mutants have been stable, whereas others, e.g. 'Macspur,' showed bud reversion (7). This, coupled with variable growth (7) and lower productivity and fruit quality than regular strains (6), has slowed acceptance of these strains. The 'Marshall McIntosh' strain (Table 1) seems to be more intensively

TABLE 1.
Some recent strains of 'McIntosh.'

Strain	Year introduced	Place of origin	Literature source
Spur-Type Growth Habit			
Greenslade[z]/Macspur	1969	B.C., Canada	2
Raikes/Morspur	1969		
Gatzke/Starkspur McIntosh	1969		
Dewar/Starkspur Ultra Mac	1969		
Wijick/Starkspur Compact Mac	1969		
Regular Growth Habit			
Marshall McIntosh	1983	Massachusetts, USA	8
Chick-A-Dee McIntosh	1984	Maine, USA	10

[z] *Strain named after orchard of origin and was renamed by the purchasing nursery (second name).*

FIGURE 1.
Historical plaque commemorating John McIntosh, who discovered the original 'McIntosh' apple seedling. This plaque, the monument, and tree marker (Fig. 3) are all at Dundela, Dundas County, Ontario. Photos taken June 14, 1988.

colored and ripens earlier than other strains, yet has good storage quality (8). As yet, there is little information about 'Chick-A-Dee McIntosh' (10).

Breeding

Plant breeders have chosen to use 'McIntosh' in their breeding programs because of its consistent yields of high quality fruit. There have been many introductions which have 'McIntosh' in their parentage (12, 14), and a number of these have achieved commercial acceptance. These include, 'Melba,' 'Spartan,' and Summerred' from Canadian breeding programs; 'Cortland,' 'Early McIntosh,' 'Empire,' and 'Macoun' from American programs; 'Merton Charm' and 'Tydeman's Red' from English programs; and 'Kyokko' from Japanese programs.

Columnar apples

A major contribution of spur-type 'McIntosh' may come from studies of the 'Wijick' mutant. This mutant is non-branching and highly spurred, and transmits these unique traits to a high proportion of its progeny (4, 5).

The 'Wijick' mutant, with its dominant gene for compact or columnar habit, now forms the basis of a columnar apple breeding program at the East Malling Research Station in England (13). Four selections from crosses made with 'Wijick' in England, 'Maypole,' 'Tuscan,' 'Trajan,' and Telamon,' are now in final evaluation for release to ornamental or commercial growers. Intensive production systems for apple demand compact, precocious cultivars.

Disease resistance

Breeding of apples for resistance to diseases, particularly apple scab, has been an important North American cooperative effort, under the Apple Breeders' Cooperative, since the 1940s. 'McIntosh' has been featured as a parent in this breeding program, and many of the releases contain it. For instance, the recently named 'McShay' apple (9) originated from the cross of 'McIntosh' x PRI 612-4 made in 1962. 'McShay' fruit are very similar to 'McIntosh' in color, light bloom and bright finish, and storage period.

Another example of 'McIntosh's' use can be found in the scab-resistant apple breeding program initiated at the Ottawa Research Station, Ontario, Canada, by L.P.S. Spangelo in 1949 (3,12). Five selections of sufficiently high quality to be named are recommended for limited trial planting. Four of these selections, 'Maefree,' 'Moira,' 'Trent,' and 'Murray,' have 'McIntosh' as one of the parents. The fruit of these selections are red, with a range of acidity and maturity dates. In addition to these selections, there are many others in the program that have 'McIntosh' as a parent and are being evaluated further.

'McIntosh'—today and tomorrow

From the time of Allan McIntosh in the late 1800s until today, we have seen, and briefly reviewed here, the tremendous impact of this apple. But what of its future? The 'McIntosh' legacy will continue through its distinctive fruit characteristics and vegetative variants which have been selected in strains and bred into new cultivars. But there is a cloud on the horizon that could challenge future 'McIntosh' production. Because 'McIntosh' fruit are susceptible to preharvest drop and softening, treatment with daminozide (Alar) has become part of the management profile. Daminozide retards preharvest drop and helps retain fruit firmness in storage. If use of daminozide was to be restricted, or lost, the old problems of dropping and softening would emerge again.

In spite of this limitation, John and Allan McIntosh can be proud of their apple seedling.

Literature cited

1) Beach, S.A. 1905. The apples of New York. Rpt. of New York Agr. Expt. Sta. 1903. Vol 1. John Wiley and Son, New York.

2) Fisher, D.V. 1970. Spur strains of 'McIntosh' discovered in British Columbia, Canada. *Fruit Var. and Hort Dig.* 24:27-32.

3) Heeney, H.B. 1981. New apple cultivars and advanced selections at the Simthfield Experimental Farm. Agriculture Canada Technical Bull. No. 2, Ministry of Supply and Services, Canada.

4) Lapins, K.O. 1969. Segregation of compact growth types in certain apple seedling progenies. *Can. J. Plant Sci.* 49:765-768.

5) Lapins, K.O. 1976. Inheritance of compact growth type in apple. *J. Amer. Soc. Hort. Sci.* 101:133-135.

6) Looney, N.E. and W.D. Lane. 1984. Spur-type growth mutants of 'McIntosh' apple: a review of their genetics, physiology and field performance. *Acta Hort.* 146:31-46.

7) Lord, W.J., R.A. Damon, Jr, and D.W. Greene. 1983. Variations in growth and productivity among 'Macspur' apple trees and growth comparisons between spur and nonspur 'McIntosh' and 'Delicious' cultivars. *Fruit Var. J.* 37:95-99.

8) Lord, W.J., W.J. Bramlage, and W.R. Autio. 1985. The 'Marshall McIntosh' apple. *Fruit Var. J.* 39:37-41.

9) Mehlenbacher, S.A., M.M. Thompson, J. Janick, E.B. Williams, F.H. Emerson, S.S. Korban, D.F. Dayton, and L.F. Hough. 1988. 'McShay' apple *HortScience* 23:1091-1092.

10) Olien, W.C. 1986. 'Chick-A-Dee McIntosh:' a new spur-type strain. *Fruit Var J.* 40:34.

11) Proctor, J.T.A. 1979. Apple cultivars grown in Canada. *Fruit Var. J.* 33:12-15.

12) Strik, B.C. and J.T.A. Proctor. 1986. Apple cultivars bred in Canada: selections from controlled crosses for commercial production. *Fruit Var. J.* 40:51-55.

13) Tobutt, K.R. 1984. Breeding columnar apple varieties at East Malling. *Scientific Horticulture* 35:72-77.

14) Upshall, W.H. 1970. 'McIntosh,' pp. 99-114. In *North American apples: varieties, rootstocks, outlook*. Michigan State Univ. Press East Lansing.

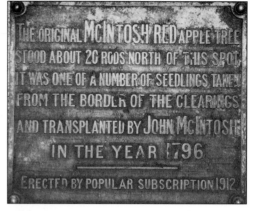

FIGURE 2.

Reproduction of the plaque on the 'McIntosh' monument.

FIGURE 3.

Engraved stone at the site of the original 'McIntosh' tree near the monument shown in Fig. 2.

Author Proctor was with the Department of Horticultural Science, University of Guelph, Guelph, Ontario, Canada, when this article originally appeared, April 1990.

The 'Bing' Sweet Cherry

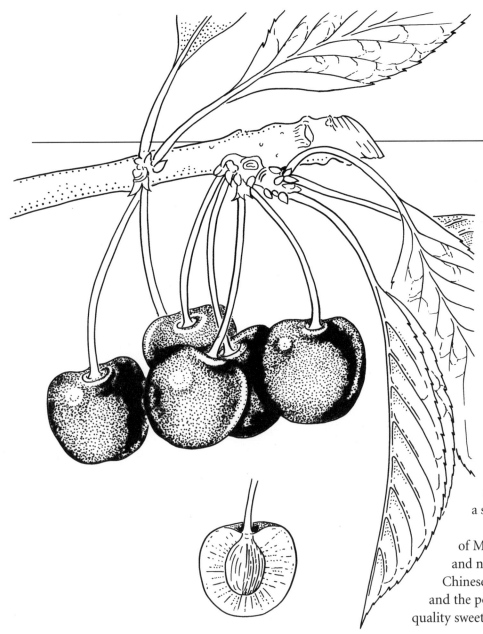

by Teryl R. Roper and Curt R. Rom

'Bing' is the predominant sweet cherry cultivar grown in the United States. Significant commercial acreage of 'Bing' is found in Washington, Oregon, and California—all western states. In 1985 Washington State had about 4,900 hectares of 'Bing' sweet cherries, representing 72% of its total cherry acreage. Hedrick (7) says of 'Bing,' ". . . few sweet cherries equal it in size and attractiveness, and none surpass it in quality, so it may be said to be as good as any of the dessert cherries."

Sweet cherries are thought to have originated in the Transcaucus region between the Black and Caspian seas. Wild cherries are found throughout Europe and Asia (7). Domestication of cherries began before recorded history. Sweet cherries belong to the genus *Prunus* and are distinguished by having fruit borne on pedicels which are one to one and one-half inches long (8). The specific name, *avium*, was adopted from their nickname bird cherry. Those who have tried to protect a sweet cherry crop from birds understand why they are called bird cherries.

'Bing' sweet cherry originated in 1875 in the nursery of Seth Lewelling of Milwaukee, Oregon, located in the Willamette Valley of western Oregon and now a suburb of Portland, OR. It is reported to have been named after a Chinese workman (7). The seed parent was 'Republican' ('Black Republican'), and the pollen parent is unknown. 'Bing' is certainly the best known, if not the best quality sweet cherry, introduced by the Lewelling nursery.

Fruits of cultivated sweet cherries can be divided into two groups according to fruit shape and texture. These groups are subdivided by juice color, which can be either light or dark. Soft, heart-shaped fruit are called heart or Gean cherries. Firm, roundish fruit are of the Bigarreau type. 'Bing' belongs to the dark-juice Bigarreau type (7, 9).

'Bing' bears fruit laterally from simple reproductive buds in the basal nodes of one-year-old wood and on short shoots or spurs on two-year-old or older wood. Terminal buds are always vegetative. Because sweet cherry produces only simple buds, when fruit are borne on the base of one-year-old wood these nodes become "blind" and produce no fruit or spur growth in future seasons. Only one bud is borne in the axil of each leaf, with one to three flowers per bud. Spurs have 3-8 lateral fruit buds, and there may be 10-20 fruiting spurs on sections of two-year-old wood. The corolla is white, and the fused base of the sepals form a hypanthium. The flower is perigynous, and the fruit is a solitary drupe (8). The leaves are large, 10-15 cm long and 5-8 cm wide (45-80 cm^2), with prominent glands which exude a gum rich in carbohydrates. Glands are located on either side of the petiole just basipetal to the leaf blade.

Bloom occurs sequentially in a basipetal direction on each limb, beginning with blossoms at the base of one-yearold wood. The best fruit come from flowers distal on the limb which are the earliest blossoms to open (10).

'Bing' is a heavy yielder (approximately 13-19 metric tonnes/hectare) with large fruit, some more than an inch in diameter. Among sweet cherries, the harvest is midseason, occurring in mid June in the Yakima valley of Washington State. Typical of sweet cherries, the interval between bloom and harvest (60 to 75 days) is the shortest of any temperate, deciduous tree fruit. 'Bing,' being a dark Bigarreau cherry, has firm, crisp flesh and dark purplish juice at maturity. The fruit is sweet (soluble solids generally 16 to 20%), and very flavorful with a high aromatic content. The fruit will attain red color without direct exposure to light on the fruit (13). Because the stems must remain in the fruit for consumer acceptance of the fresh crop, and because fruit can be easily bruised resulting in cherry pitting (4), the crop must be harvested by hand. The fresh market is based on the 'Bing' cultivar, and virtually all of the 'Bing' crop is marketed as fresh fruit. 'Bing' is, however, a popular shipping cherry, because it has firm flesh which withstands packing and shipping.

Because of its horticultural quality, 'Bing' has been used extensively in sweet cherry breeding programs (3, 5). Notable progeny include 'Rainier' and 'Chinook' (1). No commercially important sweet cherry cultivars have been selected from sports or bud mutations.

'Bing' does have some production problems. The fruit are susceptible to cracking if rains occur during harvest. Rain cracking is a major reason for the limited geographical distribution of sweet cherries. Though various spray and other control regimes have been tried, such as calcium and gibberellic acid treatments (6), rain cracking will continue to be a major limitation to growing sweet cherries.

> *The fruit will attain red color without direct exposure to light.*

After 110 years of cultivation, 'Bing' cherry still remains most highly regarded among dark sweet cherries.

Winter hardiness of wood and buds is a problem for 'Bing.' After the 1955 winter freeze in Washington State, clones of 'Bing' which exhibited the least injury were collected, and one (OB 260) has been used for propagating material. The horticultural quality of 35 clones is being investigated at Prosser, WA (12). Full bloom for 'Bing' sweet cherry is in early April, which is before the average date of the last spring frost in the Yakima Valley of Washington State (2).

Like all older cultivars of sweet cherry, 'Bing' is self-unfruitful. In addition, it is incompatible with many cultivars, requiring careful choice of a pollinizing cultivar. Unfortunately, 'Napoleon' and 'Lambert,' other major western cultivars, are incompatible with 'Bing' (13). 'Van' and 'Rainier' are frequently used as pollinizers.

Unpruned, 'Bing' sweet cherry will grow to a height of 15 meters or more, with erect, vigorous branches. Trees are typically planted at 250-370 trees per hectare at spacings from 5.5 x 7.3 m to 4.8 x 5.5 m. At present (in 1990), because of incompatibility problems or because of poor performance in cold climates, no suitable dwarfing rootstock is available for sweet cherry. Research is continuing on this problem (11). Rootstocks currently in use in the Pacific Northwest are seedlings of *Prunus avium* 'Mazzard' or *Prunus mahaleb*. New chemical growth retardants may prove useful in controlling sweet cherry tree size.

After 110 years of cultivation, 'Bing' cherry still remains most highly regarded among dark sweet cherries. It is the standard by which new cherries are judged. 'Bing' may likely remain the premier dark sweet cherry for years to come.

Authors Roper and Rom were with the Department of Horticulture and Landscape Architecture, Washington State University, Pullman, when this article originally appeared, July 1990.

Literature cited

1. Anonymous. 1960. Chinook and Rainier, new cherry varieties. Wash. Agr. Exp. Sta. Circ. 375.
2. Ballard, J. K., E. L. Proebsting, and R. B. Tukev. 1982. Critical temperatures for blossom buds, cherries. Ext. Bull. 1128. Cooperative Ext., Washington State Univ., Pullman, WA.
3. Cullinan, F. P. 1937. Improvement of stone fruits. pp. 665-748 in *1937 U.S.D.A. Yearbook of Agriculture.*
4. Facteau, T. J. 1979. Factors associated with surface pitting of sweet cherries. J. Amer. Soc. Hort. Sci. 104:706-710.
5. Fogle H. W. 1975. Cherries, pp. 348-366. *In:* J. Janick and J. N. Moore (eds). *Advances in fruit breeding.* Purdue Univ. Press, West Lafayette, Indiana.
6. Guelich, K.R. 1985. Effect of calcium, gibberellic acid, spray adjuvants, and hydrophobic agents on cracking, firmness, and other quality parameters of 'Bing' sweet cherry. M.S. Thesis, Washington State University, Pullman.
7. Hedrick, U.P. 1914. The cherries of New York. New York Agr. Exp. Stn.
8. Hedrick, U.P. 1925. *Systematic Pomology.* The Macmillan Company, New York.
9. Marshall, R. E. 1954. *Cherries and cherry products.* Interscience publishers. New York. 283 pp.
10. Patten, K. D. 1986. Factors accounting for the within tree variation of fruit quality in sweet cherries. *J. Amer. Soc. Hort. Sci.* 111:356-360.
11. Perry, R. L. 1985. Progress with cherry rootstocks. *Compact Fruit Tree.* 18:107-108.
12. Proebsting, E.L. 1986. Personal communication.
13. Westwood, M.N. 1978. *Temperate Zone Pomology.* W.H. Freeman and Co. New York.

The Australian Pistachio 'Sirora'

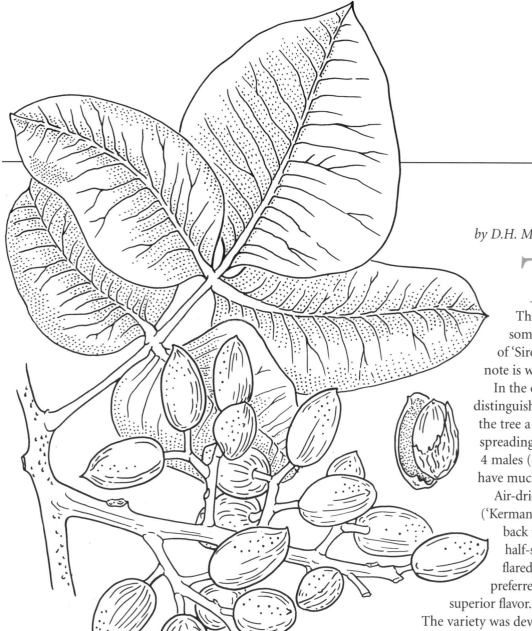

by D.H. Maggs

There are now over 300 hectares of pistachios cultivated in Australia (in 1990), about half of them 'Sirora' and about half 'Kerman,' the main Californian cultivar. The 'Kerman' nut is larger, but 'Sirora' crops more heavily, sometimes twice as much. In 1981, I published a brief description of 'Sirora' to establish the cultivar name (1); the present invited note is written to introduce it to the American industry.

In the orchard, 'Sirora' looks much like 'Kerman.' It may be distinguished by its slightly smaller and more curled leaves, which give the tree a more "mossy" appearance. Also, the tree tends to be more spreading. It is fairly late flowering, usually synchronizing with Group 4 males (2) and ripens about eight days before 'Kerman.' The fruits have much of the rosy-red coloring of 'Red Aleppo.'

Air-dried dehulled nuts typically weigh about 110 grams per 100 nuts ('Kerman' about 120-140g) and measure 18 millimeters long, 12 mm back to front, 11 mm side to side. The suture where the two half-shells meet is only a little thickened, whereas it is markedly flared in 'Kerman.' Mature kernels are green internally, a character preferred in the Old World and considered to be associated with superior flavor.

The variety was developed under the crop improvement program of the CSIRO Division of Horticulture from a selection of 'Red Aleppo.' The original seed was brought into

Mature kernels are green internally, a character preferred in the Old World and considered to be associated with superior flavor.

Author Maggs was with the Division of Horticulture, Commonwealth Scientific and Industrial Research Organisation, Merbein, Victoria, Australia, when this article originally appeared, October 1990.

Australia by Dr. D. Symon of the Waite Institute from the U.S. Department of Agriculture's field station at Chico, California. Although the varietal name 'Red Aleppo' must surely indicate an origin from Aleppo in Syria (an area known for its pistachios since Biblical times), Whitehouse (3) states that, "…the origin of the 'Red Aleppo' variety is not too clear. In 1906, it was presented… by the Rev. R.A. Fuller under the name 'Large Red Aleppo' for trial at Chico. Presumably, it is a seedling from the Turkish 'Red Aleppo' variety, the hulls of which are characteristically red." (Fuller worked for some years in Turkey.)

Bearing in mind that all pistachios are cross-pollinated, there must have been at least three outcrossings in the lineage from the tree at Aleppo to 'Sirora,' resulting in an increase in phenotypic variation—firstly, to produce the seed taken to Turkey; secondly, to produce the seed taken to Chico; and thirdly, to produce the seed taken to Merbein from which 'Sirora' was selected. At each transfer, it may reasonably be presumed that there was selection in favor of superior nuts and good orchard qualities. If the progeny raised after each transfer contained about 20 females, 'Sirora' would be the resultant of 1/20 of 1/20 of 1/20 of the total population, i.e., 1 in 8,000 females, the males being unknown locals.

This progression between the establishment of the Turkish collections in the late nineteenth century and the fruiting of the Merbein seedlings in the 1970s occupied a period lasting about 100 years. In the face of today's problems, something quicker is needed now, and, fortunately, current developments in propagation and genetic adjustment indicate the way to go.

A prerequisite for this approach is that sources of heritable variation are readily available. This implies well maintained, up-to-date variety collections. It is pleasing to note that despite stringent, parsimonious government economies, good collections are still held and utilized by the Division of Horticulture at Merbein.

Literature cited

1) Maggs, D.H. 1981. Pistachio cultivar Sirora. *J. Aust. Inst. Agric. Sci.* 47(2):109-110.

2) Maggs, D.H. 1982. An introduction to pistachio growing in Australia. Commonwealth Scientific and Industrial Research Organisation, Australia.

3) Whitehouse, W.E. 1957. The pistachio nut—a new crop for the western United States. *Econ. Bot.* 11:282-321.

The 'Gala' Apple

by A.G. White

'Gala,' one of a new breed of apple taking the world by storm, is a small to medium apple with an orange-red stripe over a cream-yellow background. The outstanding characteristics of 'Gala' fruit are its crisp, juicy flesh and its pleasant, refreshing flavor.

'Gala' owes its existence to the vision of J.H. Kidd, a New Zealand fruit grower who began making apple crosses in his Wairarapa orchard in the 1920s. Kidd had definite ideas about what he wanted of an apple, and he deplored the trend toward poor eating quality in new apple cultivars. In his view, crosses between high-yielding North American cultivars, with attractive, sweet fruit, and English cultivars with more highly flavored fruit, would produce the ideal apple.

By 1931, Kidd had selected and sold the propagation rights to 'Kidd's Orange,' a promising apple cultivar, which he had developed from a cross between 'Cox's Orange Pippin' and 'Delicious.' Encouraged by this success, he continued to make crosses into the late 1930s, producing several hundred seedlings of 'Kidd's Orange' x 'Golden Delicious.' Sadly, Mr. Kidd died before he was able to complete selection of this new generation of apples and to see his vision become a reality. A group of selections from the 'Kidd's Orange' x 'Golden Delicious' cross was included in a larger variety evaluation trial in 1952 (Fig. 1) by Dr. Don McKenzie of the DSIR, at the Havelock North Research Orchard (1). One of these, Kidd's D8, proved to be outstanding and in 1962 was named 'Gala.' McKenzie was sure that 'Gala' would be an excellent parent, and he used it in his own breeding program (2).

Commercial planting of 'Gala' began in New Zealand in 1965, and, after the initial period of debate,

FIGURE 1.

Pedigree of 'Gala' apple.

```
'Cox's Orange Pippin'  x  'Delicious'
              |
       'Kidd's Orange' x 'Golden Delicious'
              |
           'Gala'              'Royal Gala'
                               'Regal Gala'
                               'Imperial Gala'
                               'Galaxy'
```

Author White was with DSIR Fruit and Trees, Havelock North, New Zealand, when this article originally appeared, January 1991.

it was not long before the industry recognized the potential of this new cultivar and started planting it in quantity.

'Gala' has proved to be highly prone to color mutations, and many sports of it have been discovered. The discovery of the red striped sport 'Royal Gala' by another New Zealand fruit grower Bill ten Hove, on his orchard at Matamata, was a major factor in the further development of 'Gala.' 'Royal Gala' appealed to a wider range of consumer and, from the growers' point of view, it was easier to handle because it was not as susceptible to bruising as is 'Gala.' 'Royal Gala' has dominated plantings of 'Gala' since its release in 1973.

Two other notable New Zealand sports are 'Imperial Gala' and 'Regal Gala.' 'Imperial Gala' is very similar to 'Royal Gala' and was released by David Mitchell of Hastings in 1978. 'Regal Gala,' a block red 'Gala' type, was released by the Fulford brothers of Hastings in 1980. 'Galaxy' is a new sport recently patented by Ken Kiddle, also of Hastings. It has a dark red stripe on a red background color pattern.

'Gala' strains have been heavily planted in New Zealand and are projected to exceed production of 'Red Delicious' by 1994. 'Gala' is also being planted extensively in many other countries. Brazil was one of the first countries outside New Zealand to plant 'Gala' and has been exporting fruit to Europe since 1989. In France, where 'Royal Gala' is being substituted for 'Red Delicious' in generic apple advertising, significant and expanding production is taking place. More recently, considerable interest has developed in 'Gala' in the Pacific Northwest (USA).

'Gala' and its sports have confounded the experts by their rise in popularity. Although few would dispute its eating quality and attractiveness, most believed 'Gala' fruit would be too small to appeal to consumers, and the trees too difficult to manage in the orchard. It seems, however, that Kidd's theory that eating quality is the most important marketing characteristic an apple can possess has been proved correct by 'Gala.' Unfortunately, this new cultivar is being drawn into the same cycle that caused the demise of 'Delicious.' The quest for redder selections by nurseries, anxious for something that will provide them with a competitive edge, is underway without any consideration being given to eating quality. The lesson we must learn is that people buy apples to eat, not to look at, and they cannot be fooled forever.

Literature cited

1) Dinning, T. and D. Wood. 1990. 'Gala' beginnings. Mr. J.H. Kidd of New Zealand. *Pomona* (XXIII) 3:35.

2) McKenzie, D.W. 1983. Apples. In: *Plant Breeding in New Zealand*. G.S. Wratt and H.C. Smith (eds.). Butterworths, Wellington New Zealand. pp. 83-90.

The 'Jonagold' Apple

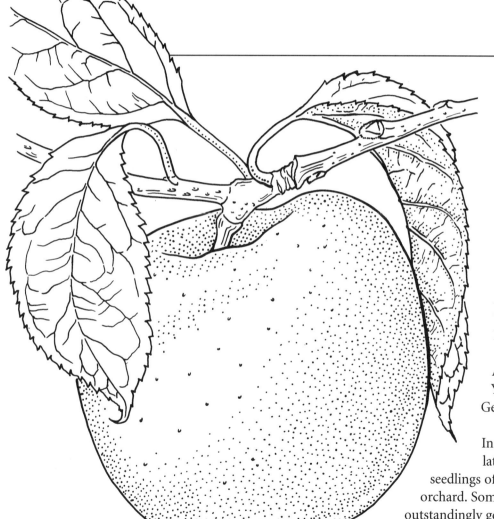

by Roger D. Way & Susan K. Brown

Out of the hundreds of apple varieties originating from apple breeding programs, 'Jonagold' is one of the few that has become extensively grown on a commercial scale by apple growers in many parts of the world. Because 'Jonagold' was introduced only two decades ago, it still has not reached its pinnacle of production. However, its planting is increasing rapidly, and there is a strong potential for even greater production in the future.

In 1988, 'Jonagold' received the Outstanding Fruit Cultivar Award from the American Society for Horticultural Science (1). This three-inch bronze medal recognizes "a modern fruit introduction having a significant impact on the fruit industry."

'Jonagold' was bred and introduced at the New York State Agricultural Experiment Station, Cornell University, Geneva, New York. It was released from the extensive apple breeding program at Geneva that started in 1895 and has been continuous to the present.

'Jonagold' was produced by conventional methods of apple breeding. In 1943, flowers on an orchard tree of 'Golden Delicious' were emasculated, and pollen of 'Jonathan' was applied to the stigmas. In 1946, 52 seedlings of this cross, growing on their own roots, were planted in a first-test orchard. Some of these seedlings bore their first fruits in 1953. One produced outstandingly good fruits and was selected for further evaluation. It was assigned the number, NY 43013-1. In 1954, this selection was budded on Malling 7 rootstocks in

'Jonagold' performs best in cooler apple-growing districts.

the nursery, and in 1956, four one-year-old nursery whips were planted into second-test orchards, where fruiting performance was further evaluated for another 10 years.

This unnamed selection was also tested by James Oakes, a commercial apple grower at Lyndonville, New York. He was enthusiastic about 'Jonagold's' performance and encouraged the experiment station to name it. 'Jonagold' was introduced in 1968 (18). 'Jonagold's' name is a combination of the names of its two parents, 'Jonathan' and 'Golden Delicious.' Commercial and home orchard growers quickly learned about its virtues, and propagating wood was distributed to nurseries. 'Jonagold' was not patented.

Statistics are not available on the number of trees nor the tons of 'Jonagold' produced in recent years in the various apple-growing regions of the world. However, 'Jonagold' is known to be an important variety in Belgium, the Netherlands, West Germany, France, Switzerland, Italy, the United Kingdom, Japan, the United States, and Canada. Schechter and Proctor (14) compiled 1986 data on 'Jonagold's' production in five countries and reported that 7% of the apples produced in the Netherlands were 'Jonagold.' Production has increased greatly since then.

In 1981, 'Jonagold' was being planted in western Europe more than any other apple cultivar, including Europe's own 'Cox's Orange Pippin' and 'Belle de Boskoop,' and also more than 'Golden Delicious,' which Europe grows so extensively (2). In 1986, the European Economic Community produced 116,000 metric tons (Belgium 50,000 MT) of 'Jonagold' apples (12).

In Belgium, 50% of all trees planted since 1977 were 'Jonagold' (5). In a presentation to the annual conference of the International Dwarf Fruit Tree Association in Toronto, Dr. H. Oberhofer, Fruit Extension Service, Lana, Italy, proclaimed, "The United States should erect a monument to the 'Jonagold' apple" (4).

'Jonagold' is a heavily yielding variety (often 1,000 bushels per acre), and it also has excellent eating quality. These attributes have contributed to the widespread interest in 'Jonagold' across Europe (15). According to Brian Lovelidge, correspondent for *The Grower* (UK) magazine in Europe, 'Jonagold' yields as well and grows almost as easily as 'Golden Delicious' but is more attractive and has far better flavor, and in the next five years, 'Jonagold' may become the prominent apple variety, replacing 'Golden Delicious' (16). Predictions indicate that by 1995, the most popular varieties in Holland will be 'Jonagold' and 'Elstar,' and 'Jonagold' will cover an estimated 22% of all apple acreage (6).

In the United States, 'Jonagold' is being tested extensively by the commercial apple growing industry in every apple growing region of the country. However, because the fruit does not have brilliant red color, its acceptance in this country has been less dramatic than it has been in Europe where more emphasis is put on good eating quality. With the recent introduction of new red strains, planting of 'Jonagold' will increase in the United States. Dr. Robert A. Norton, Washington State University, Mt. Vernon, Washington, stated that in the region west of the Cascade Mountains of Washington

State, 'Jonagold' has already become the leading apple variety (9). In Medford, Oregon, the Harry and David Company have packed 'Jonagold' in their fruit gift packages for the Fruit of the Month Club (7). Dozens of U.S. nurseries now sell trees of one or more red strains of 'Jonagold.'

The extent of commercial acceptance of 'Jonagold' in British Columbia, was assessed by Dr. Harvey A. Quamme, pomologist, Agriculture Canada Research Station, Summerland, B.C., (personal communication, April, 1988): "Apple growers in British Columbia have a strong interest in planting commercial orchards of 'Jonagold;' 50,000 trees of 'Jonagold' (a conservative estimate) have already been planted. In descending order, the three most important apple varieties planted in British Columbia in 1987 were 'McIntosh, 'Jonagold,' and 'Gala.'"

Similarly, in Japan, except for 'Tsugaru,' 'Fuji,' and 'Starking, 'Jonagold' was one of the most important varieties being planted in commercial apple orchards in the late 1980s. In October 1988, Roger Way and his wife, Mary, were invited as guests to an apple festival held by the Esashi City Apple Growers Association in northern Japan. They celebrated the 10th anniversary of the arrival of 'Jonagold' to their area. The Ways and 'Jonagold' were featured on Japanese national television. The Ways were presented with an expensive gift, an ornate Samurai Warrior's helmet, to show Japan's appreciation for this very popular new apple variety. In Japan, individual 'Jonagold' fruits sell on the retail markets for three dollars each, more than any other variety.

Fruit

The average harvest date of 'Jonagold' at Geneva is October 12, two days later than 'Delicious.' Fruits adhere well to the tree after they have reached harvest maturity. Fruits are very large, mostly 3 inches and often 3.5 inches in diameter. The shape is round-conic, nearly symmetrical and only indistinctly ribbed. The skin is 30 to 80% covered with a medium shade of red; it is only slightly striped and is slightly dull. Sometimes there is a little scarfskin. The ground color is an attractive yellow with only a slight tinge of green, which presents a pleasing overall appearance.

The fruit flesh is colored light yellow, semifirm in texture, of medium grain, crisp, and unusually juicy. It is subacid to sweet, aromatic and excellent in eating quality, having a sprightly, 'Jonathan'-like flavor. When the flesh is exposed to air, browning caused by oxidation develops slowly.

'Jonagold's' eating quality is rated by many as being the very best of any apple variety. In random tests of 600 Belgian consumers conducted by Hendrick De Coster, National Research Station, St. Truiden, 'Jonagold' always ranked first and 'Elstar' always ranked second (8). Similarly, R.A. Norton asked 19 apple experts in 9 countries to list the 10 best commercial dessert apples in the world. All 19 rated 'Jonagold' as first. Other high rating varieties were 'Gala, 'Golden Delicious,' 'Cox's Orange Pippin,' 'Fuji,' 'Elstar,' 'Empire,' and 'Red Delicious' (10).

The core of 'Jonagold' is very small relative to its large fruit size, which is a desirable attribute when fruits are being cored for processing. Smaller

All 19 apple experts, when asked to list the ten best commercial dessert apples in the world, listed 'Jonagold' as first.

cores result in fewer carpel particles which must be expensively hand trimmed from slices.

In addition to its excellent quality as fresh fruit, 'Jonagold' also has very good processing characteristics. In controlled processing tests conducted by the Food Science and Technology Department at Geneva, 'Jonagold' rated good for applesauce and fair to good for canned slices and frozen slices, while 'Northern Spy' rated excellent for all three products. In 1990, applesauce manufacturers in New York State paid $10.50 per hundred pounds for large 'Jonagold' fruits, which was a higher price than any other apple variety, except 'Northern Spy' (13).

'Jonagold' fruits have a long storage life. When stored at 31°F, they remained in a good marketable condition, without rot or shrivel, often as long as six months. The shelf life after storage is better than that of most varieties (17).

'Jonagold' performs best in the cooler apple-growing districts, such as northern Europe, northern Japan, west of the Cascade Mountains, British Columbia, and New York. But in hot districts such as eastern Washington, it generally does not perform well, because fruits often are affected by sunburn (11).

Tree

'Jonagold' trees are vigorous and somewhat above medium in size, similar to 'Golden Delicious, but not as vigorous as 'Delicious' trees. Trees are very productive, annually cropping and have excellent orchard behavior. They have no special resistances to insects or diseases. Yield comparisons in the first 10 years showed 'Jonagold' trees to be about 20% more productive (22.3 bu cumulative yield) than comparable 'McIntosh' trees (18.6 bu). Fruits are borne mainly on spurs.

'Jonagold' trees bloom in midseason along with 'Delicious,' 'Golden Delicious' is cross incompatible with 'Jonagold,' but most other diploid varieties that bloom in midseason can serve as effective pollenizers. 'Jonagold' is triploid, and therefore its pollen is not viable, and it cannot serve as a pollen source for any other variety. Because it has three sets of chromosomes, it cannot be used as a parent in a conventional apple-breeding program. Thus, it is at the end of its family line, and there is no possibility of passing its excellent qualities on to another generation, unless genetic engineering techniques can be used.

Strains

Ballard (3) stated that the Henri Fleuren Nursery in Holland tested 59 strains of 'Jonagold.' The following are some of the most widely publicized strains: 'King,' 'Wilmuta,' 'Jonagored,' 'Jomured,' 'Highwood,' 'Crowngold,' 'Nicobel,' 'Rubinstar,' 'De Coster,' 'Jonica,' 'Jonasty,' 'Jonabel,' and 'New Jonagold.' In 1989, the 'Jonagored' strain was exhibited by R. Morren-DeCoster of Belgium at the Marden Fruit Show in England, where it won the prestigious prize for the best dessert apple, and it received a challenge cup, which had never left England until then.

Definitive and comparative descriptions of these strains have not been published. These discoveries of many strains of 'Jonagold' within the brief period of two decades after introduction indicate that its skin color is genetically unstable. Thus, it has mutated several times for increased red color, and, also, strains having even less red color than the original 'Jonagold' have been observed.

All red strains have been touted as having better fruit color than the original 'Jonagold.' The relative attractiveness of color of the various strains has yet to be fully evaluated. Indeed, some do have better color, and for this reason, in the future, the original 'Jonagold' will no longer be marketed. All future propagations will be red strains. Several of them have been patented.

Literature cited

1) ASHA. 1988. Outstanding fruit cultivar award. *Amer. Soc. Hort. Sci. Newsletter* 4(10):2.

2) Bacharack, A. 1981. Interesting varieties, rootstocks discovered in Europe. *Good Fruit Grower* 32(17):33-34.

3) Ballard, J. 1990. Fruit bowl comments. *Good Fruit Grower* 41(16):31.

4) Burkhart, D.J. 1987. International Dwarf Fruit Tree Association meets at Toronto. *Good Fruit Grower* 38(9):32.

5) De Coster, J. 1986. New developments in Belgium fruit production. *Compact Fruit Tree* 19:8-21.

6) Gr. L.F.G. News. 1987. Popular apple cultivars for Dutch growers. *Great Lakes Fruit Growers News* 26(11):34.

7) Hagood, P. 1983. Several apple varieties show promise. *Good Fruit Grower* 34(11):26.

8) Holmes, S. 1987. Search for a red Jonagold busies European apple growers. *Good Fruit Grower* 38(20):83.

9) Norton, R.A. 1988. New fruit varieties from East to West. *Pomona* 21(1):41.

10) Norton, R.A. 1989. World's best commercial dessert apples. *Fruit Var. J.* 43(3):102.

11) Norton, R.L. 1988. Tree fruits. *Amer. Agriculturist,* June, p. 15, 18.

12) Oberhofer, H. 1987. Apple cultivars, rootstocks, and training systems in northern Italy. *Compact Fruit Tree* 20:36-42.

13) Pollard, K. 1990. Apple prices. Western N.Y. Apple Growers Assn. Newsletter, Fishers, NY. Oct. 9, 1990.

14) Schechter, I. and J.T.A. Proctor. 1989. 'Jonagold:' An apple for the 21st century. *Fruit Var. J.* 43(1):4-6.

15) Tukey, L.D. 1985. Jonagold is a good yielding apple cultivar. *Penn State Hort. Rev.* 34(4):2.

16) Tukey, L.D. 1987. The apple cultivar Jonagold in Europe. *Penn State Hort. Rev.* 36(1):2.

17) Way, R.D. 1971. Apple cultivars. N.Y. St. Agr. Exp. Sta. Search 1(2):27-28.

18) Way, R.D., R.L. LaBelle and J. Einset. 1968. Jonagold and Spijon: Two new apples from Geneva. N.Y. St. Agr. Exp. Sta. Circ.12.

Authors Way and Brown were with the Department of Horticultural Sciences, New York State Agricultural Experiment Station, Cornell University, Geneva, New York, when this article originally appeared, January 1986.

The 'Earliglow' Strawberry

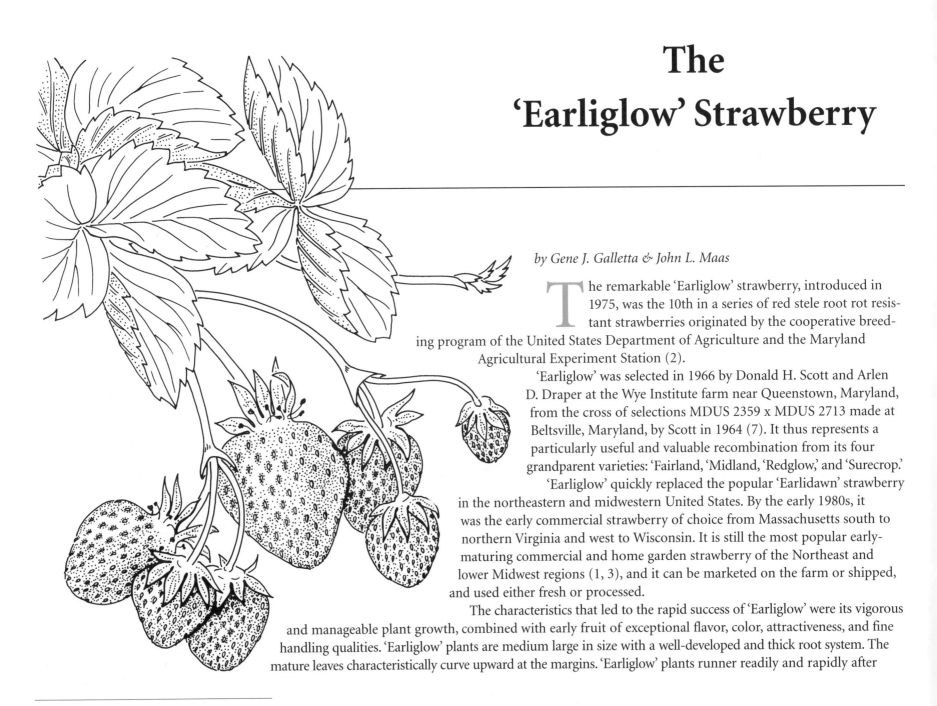

by Gene J. Galletta & John L. Maas

The remarkable 'Earliglow' strawberry, introduced in 1975, was the 10th in a series of red stele root rot resistant strawberries originated by the cooperative breeding program of the United States Department of Agriculture and the Maryland Agricultural Experiment Station (2).

'Earliglow' was selected in 1966 by Donald H. Scott and Arlen D. Draper at the Wye Institute farm near Queenstown, Maryland, from the cross of selections MDUS 2359 x MDUS 2713 made at Beltsville, Maryland, by Scott in 1964 (7). It thus represents a particularly useful and valuable recombination from its four grandparent varieties: 'Fairland,' 'Midland,' 'Redglow,' and 'Surecrop.'

'Earliglow' quickly replaced the popular 'Earlidawn' strawberry in the northeastern and midwestern United States. By the early 1980s, it was the early commercial strawberry of choice from Massachusetts south to northern Virginia and west to Wisconsin. It is still the most popular early-maturing commercial and home garden strawberry of the Northeast and lower Midwest regions (1, 3), and it can be marketed on the farm or shipped, and used either fresh or processed.

The characteristics that led to the rapid success of 'Earliglow' were its vigorous and manageable plant growth, combined with early fruit of exceptional flavor, color, attractiveness, and fine handling qualities. 'Earliglow' plants are medium large in size with a well-developed and thick root system. The mature leaves characteristically curve upward at the margins. 'Earliglow' plants runner readily and rapidly after planting to establish a moderately dense matted row or bed. 'Earliglow' also responds favorably to high density mother plant culture (ribbon row or hill culture). Plants of 'Earliglow' are resistant to 9 of the 10 U.S. races of the red stele root rotting fungus, *Phytophthora fragarieae*, and tolerant to many strains of Verticillium wilt (5, 6). Further 'Earliglow' plants are resistant to leaf scorch and leaf blight, tolerant to leaf spot, but susceptible to powdery mildew, anthracnose and angular leaf spot diseases (4).

'Earliglow' plants usually produce one or two well-developed flower clusters per crown (as compared to four or five for more recent, higher yielding cultivars). The flowering period of 'Earliglow' plants is early and dense, with the flowers producing good amounts of pollen. 'Earliglow' flowers set well, and develop into uniform, short conic fruits with a broad shoulder and a short neck. The neck is exposed at maturity by the reflexed (upcurved) cap.

'Earliglow' fruit is a bright, deep-glossy red at maturity, and its flesh is also a dark red throughout. The fruit "skin" is tough (scuff resistant) and has ample yellow "seeds," slightly raised above to flush with the fruit skin surface. 'Earliglow' fruit flesh is firm but juicy and highly aromatic. The texture of the flesh is uniformly firm in response to slicing or biting. Flowers and developing fruits of 'Earliglow' are highly resistant to the gray mold fungus *(Botrytis cinerea)* (4), but are susceptible to leather rot *(Phytophthora cactorum)*.

The celebrated flavor of 'Earliglow' fruit results from a unique combination of high soluble solids, high acidity, and high aroma. No unique volatile substances have been discovered yet in 'Earliglow' fruits. The usual consumer response, when eaten fresh, is that the berry is sweet, with a strong sharp, pleasant and balanced flavor that is distinctly strawberry-like. Moreover, 'Earliglow' can develop this flavor under a wide variety of soil and climatic conditions. 'Earliglow' fruit is frequently sought by name by consumers.

The weaker features of 'Earliglow' which are often noted are its modest fruit size and yield. Primary 'Earliglow' berries may weigh 18 to 20 grams, but later fruits can be as light as 5 to 6 grams, and average fruit weight is 8 to 8.5 grams. This average size is comparable to that of the fine 'Raritan,' 'Surecrop,' and 'Redchief' varieties, but smaller than 'Honeoye,' 'Kent,' 'Allstar,' 'Guardian,' and the recent California varieties. Similarly, 'Earliglow' may produce up to 25,000 pounds of fruit per acre, but newer varieties can produce 25,000 to 35,000 pounds per acre in the East and 40,000 to 50,000 pounds per acre in the longer ripening season of California. However, the unique combination of broad disease resistance with early, high quality fruit should ensure continued culture of the 'Earliglow' strawberry for another generation.

Additionally, 'Earliglow' has proven to be an exceptional parent, because it readily transmits its disease resistance, fruit sweetness and flavor, firmness, color, reflexed calyx, and early maturity. 'Earliglow' contributed to the parentage of the newer Canadian clones 'Annapolis,' 'Cornwallis,' and 'Cavendish,' and is a parent of many of the most promising USDA (Beltsville), NJUS, NYUS,

> *'Earliglow' fruit is frequently sought by name by consumers.*

and MNUS June-bearing selections. 'Tribute' and 'Tristar' day-neutral strawberries also have parentage which is common in part to that of 'Earliglow.' In recognition of their origination of this exceptional berry variety, Donald Scott and Arlen Draper were presented the first Outstanding Fruit Cultivar Award by the American Society for Horticultural Sciences' Fruit Breeding Working Group in 1988.

Literature cited

1) Galletta, G.J. 1989. Northeastern United States strawberry cultivars. *Fruit Var. J. 43*:40-42.

2) Galletta, G.J., A.D. Draper, and D.H. Scott. 1981. The U.S. Department of Agriculture Strawberry Breeding Program. *HortScience* 16:743-746.

3) Luby, J.J. 1989. Midwest and Plains States strawberry cultivars. *Fruit Var. J. 43*:22-31.

4) Maas, J.L. (Ed). 1984. Compendium of Strawberry Diseases. Amer. Phytopath. Soc., St. Paul. 138 pp.

5) Maas, J.L. and G.J. Galletta. 1987. Reaction of strawberry cultivars to several races of *Phytophthora fragariae*. Proc. 1987 Winter Conf. N. Amer. Strawberry Growers Assn. pp. 108-112.

6) Maas, J.L. and G.J. Galletta. 1987. Strawberry cultivar reactions to several isolates of *Verticillium*. ibid. pp. 120-130.

7) Scott, D.H. and A.D. Draper. 1975. 'Earliglow,' a new early ripening strawberry. *Fruit Var. J.* 9:67-69.

Authors Galletta and Maas were with the Fruit Laboratory, Beltsville Agricultural Research Center, USDA-ARS, PSI, Beltsville, Maryland, when this article originally appeared, July 1991.

The 'Springcrest' Peach

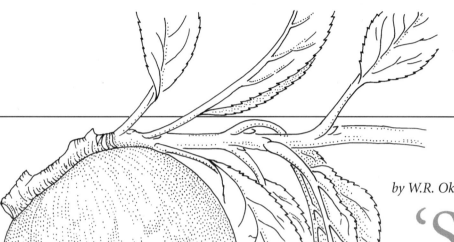

by W.R. Okie & S.C. Myers

'Springcrest' peach represented a breakthrough for commercial peach production because of its combination of early maturity, size, attractiveness, quality, firmness, and shipping ability. Commercially introduced in 1969, it has dominated the early market for nearly two decades in California and much of the world because of its superior characteristics.

'Springcrest' has been California's leading cultivar in terms of total acreage, and represented the fourth leading cultivar in total production out of 78 listed cultivars in 1990 (2). It became California's number one cultivar in 1979 and remained so until 1983, when surpassed by the late-season peach 'O'Henry.' Since 1979, 'Springcrest' and its mutations have accounted for 8 to 15% of the total fresh peach production in California. In 1988, they accounted for 15% of the freestone peach acreage (1).

It has also been a successful cultivar in Europe and in South America. In Italy, 'Springcrest' represented 13% of the total 1986 production, second only to 'Redhaven.' It is also a leading early production cultivar in France. Ironically, in Georgia where it originated, 'Springcrest' has ranked in the top 10 over the last 20 years but has not been as predominant as it has been in other areas.

'Springcrest' has also been a major source of new cultivars as a direct parent in breeding and as a source of bud sports (Table 1). In recent years, several 'Springcrest' sports, mostly slightly earlier maturing, have been gaining in importance. All of the sports in Table 1 originated in California, except for 'Starcrest' and 'Cristel,' which are from France, and 'Early Crest,' which is from Italy.

TABLE 1.

Important bud mutations and progeny of 'Springcrest' peach and their 1990 production in California. Package size is 10 kg. (CFTA,1990):

Cultivar	Type[Z]	Production (1,000 packages)
Springcrest	M	874
Maycrest	M	759
Queencrest	MM	207
Raycrest	M	135
Goldencrest	S	30
Morning Sun	M	29
Early Maycrest	MM	27
Ambercrest	S	18
Earlicrest	M	13
Ruby May	M	1
Early Crest (=San Isidoro)	M	—
Firecrest	M	—
Starcrest (=Chastar)	M	—
Cristel (=Primecrest)	M	—
Crimson Lady	S	—
Crown Princess	S	—

[Z]M = mutation, S = seedling, MM = Maycrest mutation.

Because of its major impact on commercial peach production worldwide and on the introduction of new cultivars, 'Springcrest' was awarded the Outstanding Cultivar Award in 1990 by the American Society for Horticultural Science. The medal is inscribed with the names of both V.E. Prince and J.H. Weinberger to commemorate their roles in developing the peach. Not since the days of 'Redhaven' (3), and before that 'Elberta' (4), has a single cultivar had such a significant impact on peach production.

'Springcrest,' tested as FV9-170, resulted from a cross of FV89-14 x Springtime (Figure 1) made in 1958 by the late Victor E. Prince at the USDA Horticultural Field Station in Fort Valley, Georgia (now located at Byron, Georgia). 'Springcrest' is a descendant of 'Elberta' peach, a seedling selection from Georgia that has also had a significant influence on peach production in the past (4). Parent FV89-14 is of particular interest, because it is also a parent of 'Springold,' 'Camden,' and 'Starlite,' and grandparent of 'Sunprince' (5). FV89-14 was also used in California as a parent of 'Fayette' and 'Flavorcrest' and as a grandparent of 'Flamecrest' and 'Goldcrest.' FV89-14 has produced progeny with a wide range of maturity, from May through September, but because of bacterial spot susceptibility, was never named and released. FV89-14 resulted from a cross ('Fireglow' x 'Hiley') x 'Fireglow' made by J.H. Weinberger in 1941 while he was located at Fort Valley.

'Springcrest' first fruited in 1961 and was selected in that year for testing by Prince. It was tested at some 12 state experiment stations in the Southeast. In addition, extensive testing was done in California by J.H. Weinberger at the U.S. Horticultural Field Station, Fresno, California, and in grower-cooperator trials in California. 'Springcrest's' outstanding performance in California was a major factor in its release.

'Springcrest' is less well adapted to the northern and northeastern United States because of its flower bud chill requirement of 650 hours (below 7°C). Time of bloom is approximately four days before 'Elberta' at Byron, with large-petaled, showy, light-pink blossoms which are self-fertile. Leaf glands are globose and trees are moderately vigorous but susceptible to bacterial spot [*Xanthomonas campestris* pv *pruni* (Smith) Dye].

Fruit are small to medium in size, round with a slight tip and semi-freestone when fully ripe. Flesh color is medium yellow with no red flecking. Fruit are firm but melting, medium in texture, and with a good subacid flavor. Fruit have a nonprominent suture and fine, short pubescence. 'Springcrest' has exhibited fewer split pits than most other early cultivars, but some may occur when crops are light. In California, exterior color is very attractive, having a bright red blush on 90% of the surface, over a yellow ground color. In the Southeast, 'Springcrest' fruit tend to have excessively dark red color unless trees are growing vigorously (6).

Literature cited

1) California Agricultural Statistics Service. 1988. California fruit and nut acreage 1988. Sacramento, CA.

2) California Tree Fruit Agreement. 1990. Annual report. Sacramento, CA 95865.

3) Iezzoni, A.E. 1987. The 'Redhaven' peach. *Fruit Var. J.* 41:50-52.

4) Myers, S.C., W.R. Okie, and G. Lightner. 1989. The 'Elberta' peach. *Fruit Var. J.* 43:130-138.

5) Okie, W.R., D.W. Ramming, and R. Scorza. 1985. Peach, nectarine, and other stone fruit breeding by the USDA in the last two decades. *HortScience* 20:633-641.

6) Savage, E.E. and V.E. Prince. 1972. Performance of peach cultivars in Georgia. University of Georgia Agri. Exp. Sta. Bull. 114.

FIGURE 1.

Pedigree of 'Springcrest' peach.

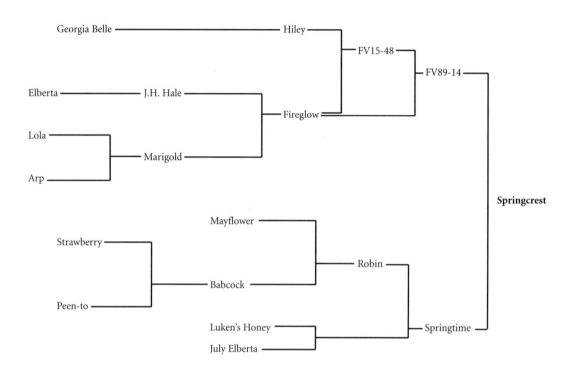

Authors Okie and Myers were with the USDA Southeastern Fruit and Tree Nut Research Laboratory, Byron, Georgia; and the Department of Extension Horticulture, University of Georgia, Athens, respectively, when this article first appeared, October 1991.

The 'Heritage' Red Raspberry

by Hugh Daubeny, Kevin Maloney, & G.R. McGregor

'Heritage' has become the world's predominant primocane fruiting red raspberry cultivar. No other primocane cultivar has ever been as widely planted, and it has become the standard by which all others are judged. It is grown in both the northern and southern hemispheres and, in the latter, is the most widely planted of any cultivar, whether it be floricane or primocane fruiting. Most of the fresh raspberry fruit seen in Northern Hemisphere markets during the winter and early spring months is 'Heritage' grown in the Southern Hemisphere, mainly in Chile. In fact, the cultivar is largely the basis of the industry in that country and has been a major factor in the dramatic increase in raspberry production there in recent years.

Elsewhere in the Southern Hemisphere, the cultivar is well suited to southern and coastal Australia and has allowed raspberry production in areas unsuited to floricane fruiting cultivars because of the lack of sufficient chilling hours. In the Northern Hemisphere, 'Heritage' has been responsible for extending the fresh market season of raspberries through September and into October and early November. In some regions such as central coastal California, production can occur even later in the fall. In Europe, the cultivar is being used successfully, on a small scale, for greenhouse forcing of potted and cool-stored plants for off-season fruit production.

There are interrelated reasons why 'Heritage' has had such an immense impact. Fruit traits which have contributed to this include its medium red color, firm texture, small drupelet, easy removal or release and the ability to remain in good condition even when overripe. The last named trait is equated with a degree of resistance to fruit rot caused by *Botrytis cinerea* Pers. ex. Fr. This and the firm texture mean the fruit can be shipped long distances and has an extended shelf life.

Plant traits contributing to the success of 'Heritage' include the production of high numbers of relatively sturdy, upright canes which may require no supports. In regions with higher temperatures during the summer months, for example, Maryland compared to New York, canes have more tendency to produce longer laterals, and this contributes to increased yields. Higher temperatures also promote earlier ripening and the chances of a greater portion of the crop ripening before unfavorable conditions occur in the fall. Under optimum cultural conditions, a substantial crop is often produced on the primocanes in the planting year.

Plants resist or show tolerance to several potentially devastating diseases, including raspberry bushy dwarf virus and the mosaic virus complex. In most production regions, plants are relatively tolerant to *Phytophthora*-incited root rots.

Although 'Heritage' will produce a spring or early summer crop on the overwintered portion of canes which did not fruit the previous season, the fruit quality of this crop is not as good as that of most of the floricane fruiting cultivars. It is thus recommended that the canes be mowed to ground level during the winter months. This has the advantage of avoiding selective cane removal and tends to promote an earlier primocane crop. Moreover, it avoids the possibility of winter injury to the canes and lessens the chances of diseases or insects affecting the canes.

'Heritage' was released in 1969 by D.K. Ourecky and G.L. Slate from Cornell University's New York State Agricultural Experiment Station breeding program at Geneva. The cultivar, which was tested as New York 696, originated from a cross between New York 463 ('Milton' x 'Cuthbert') x 'Durham' and was selected in 1960 from a population of 32 seedlings. After its release, it soon replaced older primocane fruiting cultivars such as 'September' and 'Fallred.' In the intervening years, 'Heritage' has been used extensively in various breeding programs and is a parent of several recent releases, including 'Ruby,' from the New York program and 'Redwing,' from the University of Minnesota program.

There have been attempts to find cultivars which might be superior to 'Heritage.' Of particular concern in many environments is the possibility of earlier ripening to ensure a more complete overlap with late-season floricane cultivars and, as already indicated, to ensure a greater portion of the potential crop ripening before unfavorable fall weather conditions. In addition, there is a desire for cultivars with larger fruit with a more intense raspberry flavor and a brighter color. Some of the recently released cultivars, such as 'Malling Autumn Bliss,' from the Horticulture International program in England, and 'Amity' and 'Summit,' both from the Oregon State University, United States Department of Agriculture program, as well as

> *It is recommended that the canes be mowed to ground level during the winter months.*

'Redwing' and 'Ruby' are being planted because of one or more the aforementioned traits. To date, none of the newer cultivars has proved to be as widely adapted as 'Heritage.' Therefore, it appears unlikely that any one will completely replace 'Heritage' in a particular production region, and the cultivar will remain important into the twenty-first century.

Authors Daubeny, Maloney, and McGregor were with the Agriculture Canada Research Station, Vancouver, British Columbia; the Department of Horticultural Sciences, New York State Agricultural Experiment Station, Cornell University, Geneva, New York; and the Potato Research Station, Healesville, Victoria, Australia, respectively, when this article originally appeared, January 1992.

The 'Flordaprince' Peach

by W.B. Sherman & P.M. Lyrene

Low-chill peach breeding began at the University of Florida in 1953 through the efforts of R.H. Sharpe. The main goal was to develop cultivars of fruit quality and size equal to cultivars grown in temperate zones. Low-chill adaptability was used from three sources: the feral Spanish germplasm introduced through St. Augustine, from which 'Jewel' and 'Waldo' were locally named representatives; 'Hawaiian,' an introduction from south China; and 'Okinawa,' a seed introduction from the Ryukyu Islands.

The low-chill germplasm was hybridized with temperate-zone cultivars and advanced selections from other breeders, but mainly with germplasm available through the Southeastern USDA Station in Georgia. Segregating F2 populations gave rise to selections that were intercrossed and hybridized with other high fruit quality, temperate-zone germplasm to produce advanced generations and form the basis of the current breeding program from which 'Flordaprince' was selected.

'Flordaprince' was named in honor of Victor E. Prince, the USDA Georgia peach breeder, because he cooperated for many years in the hybridization program in which crosses and backcrosses were made to incorporate low-chill genes into the best temperate-zone peach germplasm.

Reflecting on pedigree charts of the newest cultivars and selections, one can speculate on the probable source for several important traits. High fruit set ability in low-chill germplasm came from 'Okinawa.' Prior to introduction of 'Okinawa' blood into the program, most selections shed a high percentage of their flower buds prior to opening or opened weakly and set a low percentage of fruit. 'Springtime' and 'Earligold' were the main contributors of early ripening, as was 'Panamint,' the original source of the nectarine gene. 'Flordaprince' has both 'Okinawa' and 'Earligold' in its pedigree.

Ten nectarines and 21 peaches have been released by the University of Florida low-chill breeding program (1). 'Flordaprince' was the first low-chill peach of modern-day fruit quality to be cultivated worldwide in subtropical climates from Australia to Zimbabwe. It is grown outside Florida in southern Texas and California. 'Flordaprince' has been introduced into over 80 countries and territories and has been commercialized in over 20. In 1991, 'Flordaprince' received the prestigious "Outstanding Cultivar Award" from the Fruit Breeding Working Group of the American Society for Horticultural Science.

'Flordaprince,' tested as 'Fla. 5-2,' originated from complex parentage and was selected in 1975 by R.H. Sharpe and W.B. Sherman. It was named and released in 1982 for grower tests. The original description proved to be accurate: "It has fruited well where the coldest month averages 16 to 17°C," and "Trees have a winter chilling requirement of about 150 hours" (2). 'Flordaprince' peach proved that low-chill peaches with high fruit quality could be produced in subtropical climates.

Its two negative factors are that it is highly susceptible to bacterial spot (*Xanthamonas campestris*. pv. *pruni*) and, when grown in climates receiving more than about 400 chilling units, fruit tend to russet and crack at the terminal end. Susceptibility to bacterial spot has limited the use of 'Flordaprince' as a parent in the Florida program. A high percentage of 'Flordaprince' open-pollinated seedlings resemble the parent strikingly, and 'Flordaprince' strongly transmits dark red fruit stripes to many of its hybrid progeny.

'Flordaprince' has high flavor for an early-ripening peach (about 85 days from bloom to ripe). The fruit are yellow-fleshed, firm, and round, with a medium small stone. The tree tends to be upright but is easily spread by pruning. Flower buds are profuse, and flowers are showy. Early thinning is required to obtain marketable size of two-inch-plus fruit diameter.

Research with 'Flordaprince' in the subtropics has contributed much to current knowledge of peach growing in mild climates where growing seasons are long and little winter chilling occurs. These contributions include research areas such as dormancy, forced flowering and production (including out of season and biannual cropping), fruit set under high temperatures, and obvious nutritional, pests and cultural uniqueness.

Literature cited

1) Sherman, W.B. and P.M. Lyrene. 1991 Deciduous fruit cultivar development in Florida. *HortScience* 26:1-2, 91-92.

2) Sherman, W.B., P.M. Lyrene, J.A. Mortensen, and R.H. Sharpe. 1982. 'Flordaprince' peach. *HortScience* 17:998.

Authors Sherman and Lyrene were with the University of Florida, Gainesville, Florida, when this article originally appeared, April 1992.

The 'Wealthy' Apple

by Teryl R. Roper

Early settlers of the northern areas of the American Great Plains were constantly looking for fruit cultivars that would flourish in this harsh environment with short, humid summers and long, cold winters. In 1860, Peter M. Gideon of Excelsior, Minnesota, had $8.00 left after feeding and clothing his large family for the winter. He sent at least part of that money to Albert Emerson of Bangor, Maine, and received scion wood for 'Duchess,' 'Blue Pearmain,' and 'Cherry Crab' apples. He also received seed from the 'Cherry Crab' from Mr. Emerson, which he planted. From this lot of seed grew the 'Wealthy' apple. Peter Gideon said ". . . I began fruit culture in Minnesota by planting thirty named varieties of apples, a good collection of pears, plums, cherries, and quinces, a bushel of apple seed, and a peck of peach seed, and yearly for nine years planted more trees and seeds, and all kept as long as they could live in Minnesota, and at the end of ten years, all died except for one small seedling crab." (6).

'Wealthy' quickly became a popular apple in the north central states. In 1882, Suel Foster of Muscatine, Iowa, praised 'Wealthy' for its hardiness, precocity, high yields, and fruit quality in the Iowa Horticultural Society Report for that year (5). In 1902, the Experiment Station Reports for Kansas and South Dakota describe desirable apples for planting in these states. Kansas found 'Wealthy' to be too vigorous and upright, producing fruit which are "not of the best quality." (3) However, N.E. Hansen of South Dakota foundy 'Wealthy' to be a desirable apple for his state (9). 'Wealthy' was also described and recommended for cold northern areas by Beach in Volume 2 of his *Apples of New York* (1). 'Wealthy' was cited in early popular pomology books including Downing's *The Fruits and Fruit Trees of America* (4) and Hedrick's *Cyclopedia of Hardy Fruits* (10). Although it was planted primarily in North America,

> *Residents of the north central states should be grateful for the persistence of Peter Gideon, who showed that good apples can be grown successfully in this environment if proper genetic material is used.*

'Wealthy' was mentioned in at least one British publication (13), where it was also reported that 'Wealthy' was exhibited in the Imperial Fruit Show in 1925. Bultitude reports that 'Wealthy' was given the Award of Merit from the Royal Horticultural Society in 1893 (2).

Despite its early popularity, 'Wealthy' is remarkably free of synonyms and sports. 'Hydes King' was the only synonym reported by Ragan (12). 'Peter' is an apparent seedling offspring (9). French describes only one sport of 'Wealthy,' Double Red 'Wealthy,' which originated on the James G. Chase farm near Sodus, New York, in 1933 (6).

'Wealthy' fruit is medium to large sized and uniform in shape and quality. Fruit shape is round to oblate. The ground color is pale yellow to greenish with a bright red striped or splashed overcolor. The flesh is white, sometimes stained with red, tender, very juicy, somewhat aromatic, and is generally rated good to very good. The fruit ripens unevenly on the tree. If not picked several times, much of the fruit will drop before harvest. The tree is smallish in stature but grows vigorously when young (1, 2). 'Wealthy' was widely planted as a "filler" tree that was later removed when the main trees filled their allotted space, and as 'Wealthy' fruit quality declined. As trees age, fruit size diminishes, unless a good thinning program is instituted.

In 1915, 'Wealthy' production was 2.2% of the national total; by 1964, it was less than 1% of the total apple production (6). While 'Wealthy' was never a major apple on a national scale, it has been an important cultivar in the north central region. In a survey of Wisconsin orchards in 1894, E.S. Goff discovered that 'Wealthy' accounted for 15% of the trees in reporting orchards, second only to 'Duchess' (8). As late as 1960, 'Wealthy' comprised 14.2% of the apple trees found in commercial orchards in Wisconsin (14). Even in 1991, 'Wealthy' trees were 8.1% of the total number of apple trees in the northern growing area of Wisconsin, primarily Bayfield County (15). Although 'Wealthy' is not widely planted today, it is propagated and available from at least 27 commercial nurseries (11).

Residents of the north central states should be grateful for the persistence of Peter Gideon, who showed that good apples can be successfully grown in this environment if proper genetic material is used.

Literature cited

1. Beach, S.A. 1905. The apples of New York. Rpt. of New York Agr. Exp. Sta. 1903. Vol II. J.B. Lyon, Albany, New York.

2. Bultitude, J. 1983. Apples, a guide to the identification of international varieties. University of Washington Press, Seattle. p. 310.

3. Dickens, A. 1902. The experimental apple orchard. Kansas Exp. Sta. Bull. 106:41.

4. Downing, A.J. 1878. *The fruits and fruit trees of North America.* John Wiley and Sons, New York. p. 398.

5. Foster, S. 1882. Report of director of fourth district. Iowa Hort. Soc. Rpt. pp. 99-102.

6. French, A.P. 1970. Varieties of yesteryear. pp.16-30. In: *North American Apples: Varieties, Rootstocks, Outlook.* Mich. St. Univ. Press.

7. Gideon, P.M. 1899. Origin of the Wealthy apple. *American Gardening* 20:404.

8. Goff, E.S. 1895. Apple Culture. Wisconsin Agricultural Experiment Station Bulletin 45:5.

9. Hansen, N.E. 1902. A study of northwestern apples. South Dakota Station Bulletin 76:112.

10. Hedrick, U.P. 1938. *Cyclopedia of Hardy Fruits.* Macmillan, New York.

11. Isaacson, R.T. 1989. Andersen Horticultural Library's source list of plants and seeds. 214 pp. Minnesota Landscape Arboretum, Chanhassen, Minnesota.

12. Ragan, H.G. 1905. Nomenclature of the apple; A catalogue of the known varieties referred to in American publications from 1804 to 1904. U.S. Bureau of Plant Industry Bull. 56:326.

13. Taylor, H.V. 1936. *The Apples of England.* Crosby Lockwood & Son, London. pp. 253-254.

14. Wisconsin Agricultural Statistics Service. 1978. Apple trees in Wisconsin. Wisconsin Dept. of Agr., Trade and Cons. Prot. Madison, Wisconsin.

15. Wisconsin Agricultural Statistics Service. 1991. Apple and cherry trees. Wisconsin Dept. of Agr., Trade and Cons. Prot. Madison, Wisconsin.

Author Roper was with the Department of Horticulture, University of Wisconsin, Madison, when this article originally appeared, July 1992.

The 'Sharpblue' Southern Highbush Blueberry

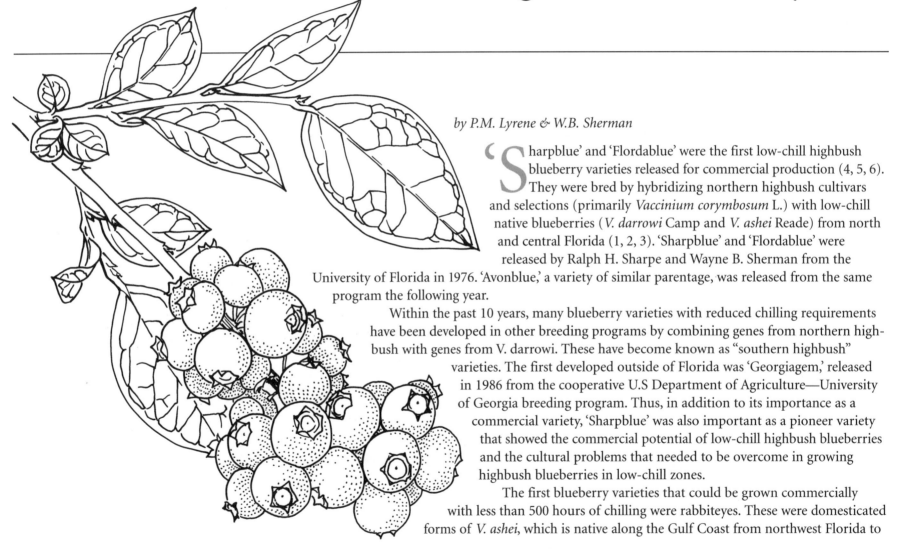

by P.M. Lyrene & W.B. Sherman

'Sharpblue' and 'Flordablue' were the first low-chill highbush blueberry varieties released for commercial production (4, 5, 6). They were bred by hybridizing northern highbush cultivars and selections (primarily *Vaccinium corymbosum* L.) with low-chill native blueberries (*V. darrowi* Camp and *V. ashei* Reade) from north and central Florida (1, 2, 3). 'Sharpblue' and 'Flordablue' were released by Ralph H. Sharpe and Wayne B. Sherman from the University of Florida in 1976. 'Avonblue,' a variety of similar parentage, was released from the same program the following year.

Within the past 10 years, many blueberry varieties with reduced chilling requirements have been developed in other breeding programs by combining genes from northern highbush with genes from V. darrowi. These have become known as "southern highbush" varieties. The first developed outside of Florida was 'Georgiagem,' released in 1986 from the cooperative U.S Department of Agriculture—University of Georgia breeding program. Thus, in addition to its importance as a commercial variety, 'Sharpblue' was also important as a pioneer variety that showed the commercial potential of low-chill highbush blueberries and the cultural problems that needed to be overcome in growing highbush blueberries in low-chill zones.

The first blueberry varieties that could be grown commercially with less than 500 hours of chilling were rabbiteyes. These were domesticated forms of *V. ashei*, which is native along the Gulf Coast from northwest Florida to southeast Georgia. Although rabbiteye blueberries were highly productive and well adapted in north Florida and in the coastal states from east Texas to North Carolina, they had two major disadvantages that led Ralph Sharpe to begin breeding low-chill highbush blueberries, beginning about 1948.

First, rabbiteyes ripened late in the season compared to highbush. Highbush blueberries produced in North Carolina ripened before rabbiteye blueberries grown in Florida. Second, fruit set on rabbiteye varieties was not reliable south of Gainesville in the northern Florida peninsula, and commercial production was not feasible south of Orlando.

Sharpe noted that the well-adapted evergreen native blueberry of the Florida peninsula, *V. darrowi*, could be hybridized much more readily with the northern highbush blueberry than with the rabbiteye. Several factors indicated good possibilities for the development of low-chill, early ripening, heat-adapted highbush blueberries. *V. darrowi* was highly vigorous and well adapted in Florida. Improved northern highbush cultivars from Michigan and New Jersey were a potential source of genes for high fruit quality, high yields, and early-season ripening. Fertile hybrids could be obtained between these two gene pools. A major breeding program to combine the best features of *V. darrowi* and the northern highbush cultivars was begun by Sharpe in 1950. By 1975, approximately 150,000 hybrid seedlings had been grown and evaluated.

The following description of the origin of 'Sharpblue' and 'Flordablue' is taken from Sharpe and Sherman (6): "A diploid species, *V. darrowi*, native to central Florida, was chosen for adaptability and crossed with the hexaploid species, *V. ashei,* native to northwest Florida. This cross resulted in five tetraploid seedlings from approximately 7,500 hand pollinations. *V. darrowi* was also crossed with northern highbush cultivars, which resulted in 31 tetraploid hybrids from approximately 1,600 hand pollinations. These 31 hybrids were apparently a result of unreduced pollen in the diploid species, *V. darrowi*. Further crosses were made by intercrossing the above original tetraploid hybrids and by backcrossing them to northern highbush cultivars. Seedlings with better fruit quality and adaptability to Florida's climate were selected, and hybridization continued. 'Flordablue' and 'Sharpblue' blueberries originated at Gainesville, Florida, from 1964 to 1965 crosses of Florida selections with complex parentage."

As of 1989, approximately 700 acres of southern highbush had been planted in Florida, and 90% of these plants were 'Sharpblue.' In addition, 400 acres of 'Sharpblue' have been planted in eastern Australia. The low-chilling requirement of 'Sharpblue' limits its usefulness outside of Florida in the United States, but the variety could be useful in certain highland areas of the tropics. In spite of the fact that much of the 'Sharpblue' pedigree traces back to lowbush *V. darrowi,* 'Sharpblue' has the highbush blueberry growth habit. It has large, dark-green leaves which tend to remain evergreen in Florida if the plants are fertilized and watered heavily during the fall and if leaf diseases are not severe. However, growers normally reduce fertilization and watering in the

In spite of the fact that much of the 'Sharpblue' pedigree traces back to lowbush V. darrowi, *'Sharpblue' has the highbush blueberry growth habit.*

fall to increase plant dormancy and reduce fall and winter flowering. 'Sharpblue' roots readily from softwood cuttings and is easy to grow in the nursery.

The ability of 'Sharpblue' to set fruit after self-pollination has been studied, and the results have varied. The number of bees available for pollination and temperatures during pollination may greatly affect self-fruitfulness in 'Sharpblue.' Observations in commercial blueberry fields in north and central Florida suggest that 'Sharpblue' should not be planted in solid blocks without a companion variety for cross pollination. Fruit sets, fruit size, yields, and earliness all appear to be reduced in solid blocks of 'Sharpblue,' compared to blocks in which 'Sharpblue' is interplanted with cross pollinators such as 'Flordablue' and 'Misty.'

'Sharpblue' normally begins flowering in late January to mid-February in the Gainesville area in north central Florida. Overhead irrigation is necessary to protect against late freezes in this area. First commercial harvest is normally during the last week in April. When grown 150 miles farther south, in the Sebring area of south central Florida, which averages about 150 chill hours per winter, 'Sharpblue' normally flowers during January, February, and early March. Commercial harvest begins in late March on farms where the early flowers have been protected from freezes, and continues into late May. In Coffs Harbor, N.S.W. Australia, 'Sharpblue' continues to flower and fruit throughout the year, and some fruit can be harvested in every month of the year.

'Sharpblue' berries are medium to large, averaging over two grams per berry in well cross-pollinated field plantings. The color of berries on the bush is light blue, but berries tend to become dark blue during harvest and packing as the surface wax is disturbed. 'Sharpblue' berries have excellent flavor and texture. At the time of its release, 'Sharpblue' was not recommended for planting as a shipping berry because of limited firmness and because the scar has a tendency to tear upon picking. Growers found, however, that dry weather, which is normal before and during the 'Sharpblue' harvest season in Florida, together with hand-harvesting and careful postharvest handling, enabled them to ship the fruit successfully on the fresh market.

Thus, 'Sharpblue' added at least one month to the season of availability of fresh blueberries on the world market, since 'Sharpblue' berries can be shipped from Florida at least one month earlier than both northern highbush blueberries from southeastern North Carolina and rabbiteyes from north Florida and southeast Georgia.

As the first commercial highbush variety with a chilling requirement of less than 800 hours, 'Sharpblue' has been important in demonstrating the possibilities of low-chill highbush blueberries. It has also been important in showing what characteristics future low-chill blueberry varieties will need to be successful in the southeastern United States. Although total acreage of 'Sharpblue' in Florida is not great, there are many small and experimental plantings, and what has been learned from these plantings is laying the groundwork for larger plantings of 'Sharpblue' or other low-chill highbush varieties in the future.

Literature cited

1) Sharpe R.H. 1954. Horticulture development of Florida blueberries. Proc. Fla. State Hort. Soc. 66:188-190.

2) Sharpe, R.H. and G.M. Darrow. 1959. Breeding blueberries for the Florida climate. Proc. Fla. State Hort. Soc. 72:308-311.

3) Sharpe, R.H. and W.B. Sherman. 1971. Breeding blueberries for low chilling requirement. *HortScience* 6:145-147.

4) Sharpe, R.H. and W.B. Sherman. 1976. 'Sharpblue' blueberry. *HortScience* 11:65.

5) Sharpe, R.H. and W.B. Sherman. 1976. 'Flordablue' blueberry. *HortScience* 11:64-65.

6) Sharpe, R.H. and W.B. Sherman. 1976. 'Flordablue' and 'Sharpblue,' two new blueberries for central Florida. Univ. Fla. Agri. Exp. Sta. Circ. S-240.

Authors Lyrene and Sherman were with the University of Florida, Gainesville, Florida, when this article originally appeared, October 1992.

The 'Meeker' Red Raspberry

by Patrick P. Moore & Hugh A. Daubeny

'Meeker' has replaced its parent, 'Willamette,' as the most widely planted red raspberry cultivar in the Pacific Northwest, which is the major producing region for the crop in North America. Currently, 'Meeker' occupies more than 60% of the plantings in the region, which includes western parts of Oregon and Washington and southwestern British Columbia. The importance of the cultivar is increasing since it is being used for most new plantings. It has also become an important cultivar in southern Chile, where there has been a dramatic increase in red raspberry production in recent years.

'Meeker' originated in the Washington State University red raspberry breeding program from the 1950 cross of 'Willamette' x 'Cuthbert' made by the late C.D. Schwartze. It was selected in 1953 and tested as WSU 408 prior to its release in 1967 (15). The name was chosen in honor of Ezra Meeker, a pioneer of the Puyallup Valley where the cultivar was selected and first tested.

As a selection, 'Meeker' was described by C.D. Schwartze as having long clusters of fruit, tall canes, long well-spaced laterals, and thus the fruit is easily seen. The habit was likened to one of its parents, 'Cuthbert.' The fruit was very late ripening, large, conic, smooth and regular, coherent, with drupelets of medium size, uniform, and medium firm. The fruit picked and handled well. The flavor was described as good, but inferior to that of the then important cultivar 'Washington.' The fruit was highly rated as a frozen product with good flavor, color, and shape retention. Compared to fruit of the cultivars grown at the time of 'Meeker's' release, such as 'Willamette,' 'Washington,' and 'Sumner,' 'Meeker' fruit had a larger number of smaller-sized drupelets. Also, its fruit was higher in soluble solids and lower in acidity (17). This gave a high soluble solids to acidity ratio and perception of 'Meeker' fruit as sweet. Preserves made from 'Meeker' fruit had good color and flavor. For processing, the color of the fruit was brighter and did not darken to the same extent as that of 'Willamette.'

Because of its very long laterals, there was concern at the time of release of the suitability of 'Meeker' for harvesting (15). However, this has not been a problem since the laterals are strong and well attached to the cane, and the fruit releases readily from the receptacle. In fact, 'Meeker' has proven as well adapted to machine harvesting as 'Willamette,' which is used extensively throughout the Pacific Northwest for fruit destined for processing uses. One reason that 'Meeker' has replaced 'Willamette' is because of higher yields along with larger fruit size. In addition, 'Meeker' fruit is more versatile because its lighter and brighter color and firmer texture is better suited to fresh market use. At the same time, the fruit is well suited to all processing uses.

'Meeker' has some field tolerance to *Phytophthora*-incited root rot, and this is still another reason why it became an important cultivar in the Pacific Northwest. However, on some sites, 'Meeker' appears to be susceptible, and this is of concern. Tolerance to root rot is becoming ever more important with the increasing spread and severity of root rot throughout the Pacific Northwest. It is resistant to both cane Botrytis (*Botrytis cinerea* Pers. ex. Fr.) and cane spot (*Elsinoe veneta* Burkh.), but susceptible to spur blight [*Didymella applanata* (Niessl) Sacc.] (5, 9, 12). It is resistant to yellow rust [*Phragmidium rubi-idaei* (D.C.) Karst.] (1, 2) and powdery mildew [*Sphaerotheca macularis* (Fr.)] (Daubeny, unpublished). It is susceptible to the common isolate of the pollen-transmitted raspberry bushy dwarf virus, with infected plants showing reduced yields. (7). However, it does not appear to become as readily infected as some other cultivars. It is susceptible to *Amphorophora agathonica* Hottes, the aphid vector of the raspberry mosaic virus complex. However, the cultivar may have some tolerance to the virus complex, since symptoms are seldom seen.

The fruit is less susceptible than that of 'Willamette' to both pre- and postharvest fruit rots, the former caused mostly by *B. cinerea* and the latter by both *B. cinerea* and *Rhizophus* species (4). Reduced susceptibility to postharvest fruit rot is a contributing factor to the extended shelf life shown by 'Meeker' fruit (11). 'Meeker' fruit also has lighter and brighter color and greater firmness compared to 'Willamette.' In British Columbia in 1990, 'Meeker' appeared less susceptible to sun damage than 'Willamette,' 'Chilliwack,' 'Comox,' 'Chilcotin,' and 'Skeena,' and also less susceptible than cultivars and selections from breeding programs in Britain.

Although considered large when released, 'Meeker' fruit is now considered medium or small in size compared to that of the recently released Pacific Northwest cultivars 'Comox,' 'Centennial,' and 'Tulameen' (5, 6, 14). It is still

Although considered large when released, 'Meeker' fruit is now considered medium or small in size compared to the recently released Pacific Northwest cultivars 'Comox,' 'Centennial,' and 'Tulameen.'

considered to be high yielding, though both 'Comox' and 'Tulameen' outyield it in British Columbia (6).

'Meeker' is not as cold hardy as some other Pacific Northwest cultivars (10). It was severely damaged at Abbotsford, British Columbia, by unusually low (-14°C) late fall temperatures (5). In laboratory freezing tests, 'Meeker' cold acclimated more slowly in the fall than 'Chilliwack,' 'Comox,' and 'Skeena.' However, in the spring, 'Meeker' was slower to lose freeze tolerance than other cultivars (10). 'Meeker' may escape cold damage when 'Willamette' is injured in years with cold weather after a warm fall. Under these conditions, 'Willamette' will break bud on primocanes and produce fall fruit, whereas 'Meeker' does not produce fall fruit.

'Meeker' is slower to break bud than many other cultivars and appears to have a relatively high chilling requirement (11). In warm climates, such as southwestern Australia and Israel, the cultivar is susceptible to blind bud which is associated with sub-optimum chilling, especially when stress is present (13, 16). Because of the aforementioned limitations, 'Meeker' is not suited to regions with particularly low winter temperatures or to regions with warm winters. However, 'Meeker' fruit showed higher drupelet set, compared to three other Pacific Northwest cultivars, 'Chilcotin,' 'Haida,' and 'Willamette,' under relatively cool growing conditions in Scotland (3).

Like its parent, 'Willamette,' 'Meeker' has remained genetically stable compared to some other cultivars (8). Of particular importance is the fact that mutations affecting fertility, expressed as crumbly fruit, have not been observed.

Also like 'Willamette,' 'Meeker' plants will establish and grow more rapidly than those of some other cultivars.(8). Under optimum cultural conditions the cultivar produces a substantial crop the year after planting.

'Meeker' has been used and continues to be used extensively in many breeding programs. It is a parent of two recent cultivar releases, 'Centennial' ('Meeker' x 'Skeena'), from Washington, and of 'Meco' ('Meeker' x 'Rose de Cote d'Or'), from France.

In 1992, 'Meeker' was awarded an Outstanding Fruit Cultivar Award by the American Society of Horticultural Sciences. 'Meeker' is now the standard for productivity and fruit quality by which newer cultivars for the Pacific Northwest are judged. Any potential replacement for 'Meeker' will have to be more productive and have larger firmer fruit with otherwise similar qualities. It will need higher levels of resistance to root rots and will benefit from resistance to several other diseases and pests and be adapted to a wider range of environmental conditions.

Literature cited

1) Anthony, B.M., R.C. Shattock, and B. Williamson. 1985. Interaction of red raspberry cultivars with isolates of *Phragmidium rubi-idaei*. *Plant Pathol.* 34:521-527.

2) Anthony, B.M., B. Williamson, D.L. Jennings, and R.C. Shattock. 1986. Inheritance of resistance to yellow rust *(Phragmidium rubi-idaei)* in red raspberry. *Ann. Appl. Biol.* 109:365-374.

3) Dale, A. and H.A. Daubeny. 1985. Genotype-environment interactions involving British and Pacific Northwest red raspberry cultivars. *HortScience* 20:68-69.

4) Daubeny, H.A. 1986. The British Columbia raspberry breeding program since 1980. *Acta Hort.* 183:47-58.

5) Daubeny, H.A. 1987. 'Chilliwack' and 'Comox' red raspberries. *HortScience* 22:1343-1345.

6) Daubeny, H.A. and A. Anderson. 1991. 'Tulameen' red raspberry. *HortScience* 26:1336-1338.

7) Daubeny, H.A., J.A. Freeman, and R. Stace-Smith. 1982. Effects of raspberry bushy dwarf virus on yield and cone growth in susceptible red raspberry cultivars. *HortScience* 17:645-647.

8) Daubeny, H.A., F.J. Lawrence, and G.R. McGregor. 1989. 'Willamette' red raspberry. *Fruit Var. J.* 43:46-48.

9) Daubeny, H.A. and H.S. Pepin. 1974. Susceptibility variations to spur blight *(Didymella applanata)* among red raspberry cultivars and selections. *Plant Dis. Rptr.* 58:1024-1027.

10) Hummel, R.L. and P.P. Moore. 1990. Seasonal variation in freezing tolerance of red raspberry clones in the Pacific Northwest. *HortScience.* 25:1087.

11) Jennings, D.L., H.A. Daubeny, and J.N. Moore. 1991. Blackberries and raspberries *(Rubus)*. pp. 331-389. *In:* J.N. Moore and J.R. Ballington, Jr. (Eds.). Genetic resources of temperate fruit and nut crops. International Society of Horticultural Science, Wageningen, the Netherlands.

12) Jennings, D.L. and G.R. McGregor. 1988. Resistance to cane spot *(Elsinoe veneta)* in the red raspberry and its relationship to resistance to yellow rust *(Phragmidium rubi-idaei)*. *Emphytica* 37:173-180.

13) Jennings, D.L., G.R. McGregor, J.A. Wong, and C.E. Young. 1986. Bud suppression ('Blind bud') in raspberries. *Acta Hort.* 183:285-289.

14) Moore, P.P., T.M. Sjulin, B.H. Barritt, and H.A. Daubeny. 1989. 'Centennial' red raspberry. *HortScience* 25:484-485.

15) Schwartze, C.D. 1967. A look back and ahead in raspberry breeding. *Proc. West. Wash. Hort. Assoc.* 57:79-82.

16) Snir, I. 1986. Growing raspberries under subtropical conditions. *Acta Hort.* 183:183-190.

17) Wolford, E.R. 1967. Observations on the Meeker raspberry for freezing and preserving *Proc. West. Wash. Hort. Assoc.* 57:69-70.

Authors Moore and Daubeny were with Washington State University Puyallup Research and Extension Center, Puyallup, Washington; and Agriculture Canada Research Station, Vancouver, British Columbia, Canada, respectively, when this article originally appeared, January 1993.

The 'Empire' Apple

by M. Derkacz, D.C. Elfving, & C.G. Forshey

In 1945, unusually warm spring weather, a very early bloom in mid-April, and postbloom frosts eliminated virtually all crosses made for the New York State Agricultural Experiment Station apple breeding program in Geneva, New York. The apple breeding material for 1945 was comprised of seed collections from isolated commercial orchards in the Hudson Valley containing only the desired cultivar pairs for crosses. One such collection contained 'McIntosh' seeds from an orchard of mature 'McIntosh' and 'Delicious' trees near Claverack, New York. A seedling tree from this collection was selected as NY45500-5 in 1954. This seedling was named 'Empire' in 1966.

Many apple cultivars widely grown today, such as 'Delicious,' 'Golden Delicious,' 'McIntosh,' and 'Granny Smith,' were discovered by chance (3, 4, 6, 13). Today, most apple cultivars originate from controlled crosses in breeding programs. Occasionally, however, even this process can have its improbable combinations of chance and good luck. Such is the case with the 'Empire' apple.

By 1945, apple breeding had been underway at the New York State Agricultural Experiment Station in Geneva for over half a century (5). In 1945, unusually warm temperatures reaching as high as 31°C occurred in late March and early April (7). Daily minimum temperatures during that period were also unusually high, falling to freezing only occasionally. Apple orchards in western New York began to bloom by April 14, about one month earlier than normal (7). Several frosts following this early bloom eliminated virtually the entire apple crop in western New York (7). The apple breeders at Geneva made their usual crosses during this early bloom, but the late April freezes left virtually no apples from which seedlings could be obtained (1).

Dr. A.J. Heinicke, head of the department of pomology and director of the experiment station, identified the solution to this dilemma. The apple breeding program emphasized crosses between commercially important cultivars to obtain late-maturing apples (1). One frequently-used cross was 'McIntosh' and 'Delicious.' Dr. Heinicke suggested that the breeders locate relatively isolated, cropping commercial orchards containing only any two cultivars normally used for crosses (C.G. Forshey, personal communication with Heinicke). Since the apples from such orchards would most likely represent natural cross-pollinations between those two cultivars, the necessary seedling material for the 1945 breeding program could be obtained.

Since almost all the New York State apple crop in 1945 was in the Hudson Valley, orchards meeting the necessary criteria were selected there for seed collection. One such orchard, owned by Mr. Asrow Miller, was located south of Claverack, New York (21, 22 C.G. Forshey, personal communication with Dr. J. Einset). This relatively isolated orchard consisted of mature 'McIntosh' and 'Delicious' trees only. At harvest, 4,035 seeds were extracted from 'McIntosh' apples in the Miller orchard and were sent to Geneva (22). Seedlings (1,199) originating from that orchard were planted in the station test orchard in spring 1947 (22).

In 1954, a Miller orchard seedling having desirable, 'McIntosh'-type fruit characteristics was assigned the selection number NY45500-5 (22). The identity of the person who selected NY45500-5 is unrecorded. Mr. Leo G. Klein, research associate at Geneva from 1949 until his death in 1962, was actively involved in the apple breeding program, wrote extensively on apple cultivars and selections during his career (e.g., 5, 8, 9, 10, 11, 12, 21), and may well have made the selection. Curiously, although he devoted considerable attention to alternatives for 'McIntosh,' Klein never mentioned NY45500-5 in any of his written material.

Dr. Roger D. Way, who also joined the pomology department in 1949, took responsibility for the apple breeding program in 1962, the year Mr. Klein died (19). By 1965, Dr. Way had recognized the potential of NY45500-5.

NY45500-5 was first described in writing in two listings of apple cultivar characteristics authored by Dr. Way and dated January 8, 1965 (14, 15). He also mentioned NY45500-5 in his presentation on apple cultivars at the annual meeting of the New York State Horticultural Society a few weeks later (16). By December 1965, Dr. Way had described NY45500-5 briefly in three published articles (16, 17, 18) and featured it on the cover of a station circular (19).

In late 1965, the decision was made to name and release NY45500-5. Suggestions for a name were solicited from the fruit industry (20). At least 104 suggestions were received from growers, packers, and others (2). The name 'Empire' was included among those suggested, but the originator(s) of the name are unrecorded. The final list of candidate

At harvest, 4,035 seeds were extracted from 'McIntosh' apples in the Miller orchard and were sent to Geneva, New York.

A History of Fruit Varieties

names included the following: 'Delight,' 'Delmac,' 'Empire,' 'Joy,' 'Nymac,' 'Polymac,' 'Red Jacket,' 'Sparkle,' 'Sprite,' and 'Tasty' (2).

The shippers expressed a strong preference for the name 'Empire' (Forshey, personal communication with Dr. J. Einset). A majority of those polled also favored this name. NY45500-5 was officially released by the New York State Agricultural Experiment Station as 'Empire' in September 1966 (22).

It is impossible to calculate the odds against the one seed with the genetic traits of the 'Empire' apple being formed in 1945 and also being in the right place at the right time to be collected, sent to Geneva, and grown in a test orchard. In addition to the unusual weather and crop loss at Geneva in 1945, several people had important roles in the discovery and development of the 'Empire' apple. We are fortunate that these unlikely events occurred, because the 'Empire' apple represents a worthy addition to the cultivars available to growers and consumers.

Literature cited

1) Anonymous, 1946. Annual report of the Pomology Department. Ann. Rpt. N.Y. State Agr. Expt. Sta. 65:51.

2) Anonymous, 1966. 104 names suggested for NY45500-5. N.Y. State Hort. Soc. Newsletter 22(2):2.

3) Ballard, J.K. 1981. 'Granny Smith'—An important apple for the Pacific Northwest. Washington State Univ. Coop. Ext. Bul.

4) Baugher, T.A. and S.H. Blizzard. 1987. 'Golden Delicious' apple—famous West Virginian known around the world. *Fruit Var. J.* 41:130-132.

5) Einset, J. and L.G. Klein. 1960. Apple improvement—early history and current aims. Farm Res. 26:4-5.

6) Fear, C.D. and P.A. Domoto. 1986. The 'Delicious' apple. *Fruit Var. J.* 40:2-4.

7) Hoffman, M.B. 1946. Why some trees and varieties set fruit and others did not in 1945. *Proc. N.Y. State Hort. Soc.* 91:96-101.

8) Klein, L.G. 1958. New varieties. *Amer. Fruit Grower* 78:58-59.

9) Klein, L.G. 1960. Notes on newer apple varieties. *Proc. 66th Ann. Mtg. Mass. Fruit Growers' Assn.* 66:105-112

10) Klein, L.G. 1962. Keep your eye on these varieties. *Amer. Fruit Grower* 82(7):10-11 18.

11) Klein, L.G. and J. Einset. 1957. Newer dessert apples. *Fruit Var. Hort. Dig.* 12:18.

12) Klein, L.G. and R.D. Way. 1961. Newer apple varieties. Farm Res. 27(4):11.

13) Proctor, J.T.A. 1990. The 'McIntosh' apple. *Fruit Var. J.* 44:50-53.

14) Way, R.D. 1965a. Tree characteristics of some apple varieties. N.Y. State Agr. Exp. Sta. Mimeo.

15) Way, R.D. 1965b. Fruit characteristics of some apple varieties. Proc. N.Y. State Agr. Expt. Sta. Mimeo.

16) Way, R.D. 1965c. Description and evaluation of apple varieties. *Proc. N.Y. State Hort. Soc.* 110:157-161.

17) Way, R.D. 1965d. Newest New York apple varieties. *Fruit Var. Hort. Dig.* 19:19.

18) Way, R.D. 1965e. Description and evaluation of apple varieties. *Eastern Fruit Grower* 28(5):12-14.

19) Way, R.D. 1965f. Tree and fruit characteristics of some standard and new apple varieties. N.Y. State Agr. Exp. Sta. Res. Circ. 3.

20) Way, R.D. 1966. Wanted: A name for the new apple NY45500-5. N.Y. State Hort. Soc. Newsletter 22(2):2.

21) Way, R.D. 1971. Apple cultivars introduced by the New York State Agricultural Experiment Station 1914-1968. Search—Agriculture, Pomology, No. 1, N.Y. State Agr. Expt. Sta., Geneva.

22) Way, R.D. and J. Einset. 1966. Introducing 'Empire'—A new dessert apple. Farm Res. 32(2):8.

Authors Elfving and Forshey were with the Horticultural Research Institute of Ontario, Vineland Station; and Hudson Valley Laboratory, New York State Agricultural Experiment Station, Highland, respectively; and author Derkacz was a graduate student, when this article originally appeared, April 1993. It was awarded a U.P. Hedrick First Place Award in 1992 by the American Pomological Society.

The 'Marsh' Grapefruit

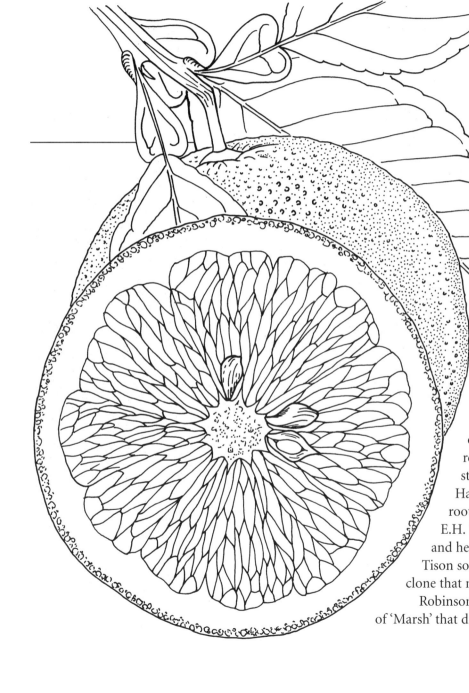

by Federick G. Gmitter, Jr.

All of the commercially significant citrus scion species are of Old World origin, with the notable exception of the grapefruit (*Citrus x paradisi* Macf.), which presumably originated in the Caribbean region in the seventeenth century (2, 3, 9, 14). Grapefruit germplasm, in the form of seed or young plants, was introduced to Florida by Count Odette Phillippi around 1823 (12).

It was from this original introduction that all of the major grapefruit cultivars are believed to be derived. Seeds and budwood were distributed throughout Florida from Phillippi's grove, and the cultivar 'Marsh' originated from one of these propagants.

Two somewhat conflicting stories have been published regarding the origin of the 'Marsh' grapefruit. One was presented by H.J. Webber (17) in reiteration of an account by E.H. Tison in the *Los Angeles Times* (16). This story holds that 'Marsh' originated in 1879 or 1880 on the farm of Mr. John Hancock, near Lakeland, Florida, as a bud sprout from a plowed up, broken root piece of another grapefruit tree that produced white-fleshed seedy fruit. E.H. Tison, a nurseryman, became aware of this seedless grapefruit around 1886, and he received seeds and budwood from Hancock for propagation in his nursery. Tison sold his nursery in 1890 to C.M. Marsh, who advertised and promoted this clone that now bears his name.

Robinson (11) published an alternative account of the details surrounding the origin of 'Marsh' that disputes some of the specifics of the former version. Robinson asserts that

'Marsh' originated as one of three grapefruit seedlings planted by Mrs. Rushing on a farm later purchased by William Hancock (John Hancock's father) in 1862. Robinson interviewed surviving family members, including John Hancock's brother, who claimed that the original tree was a large seedling, more than 30 years old, when buds were first collected for propagation, and not the much younger tree described in Webber's account (11, 17). The question surrounding the exact details of the origin of 'Marsh' grapefruit, specifically whether it originated as a seedling or a bud sprout from a root piece, will likely remain unanswered.

What is certain is that the 'Marsh' seedless grapefruit originated near Lakeland, Florida, in the late nineteenth century. Several people recognized the unique seedless characteristic that distinguished this clone from the standard seeded grapefruit, and they began to propagate it. C.M. Marsh was the individual most responsible for promoting the new cultivar after he purchased all of the existing nursery stock in the Lakeland area, so it seems appropriate that it bears his name.

'Marsh' grapefruit trees are vigorous, large, and capable of producing good crops annually. The attractive fruit possess smooth, thin, and shiny yellow rinds when grown in suitable subtropical environments. The flesh is pale buff, tender, and quite juicy. Minimum fruit maturity standards can be met by October in Florida, but fruit can remain on the tree as long as the following March or April. This on-tree storage ability provides growers with flexibility when making harvest and marketing decisions.

The flavor of 'Marsh' fruit is typical of, but less pronounced than, the seedy 'Duncan' grapefruit. It was the characteristic of producing very few seeds per fruit (usually three or less) that attracted attention to 'Marsh' and served as the basis for its early promotion; this remains a factor in the continued popularity of 'Marsh' with consumers. The good tolerance of storage conditions (on-tree and postharvest) and shipping ability of 'Marsh' fruit has allowed market expansion and exploitation. The vigor and productivity of the tree—especially when compared with several of the newer, pigmented grapefruit cultivars—has encouraged continued grower interest and acceptance of 'Marsh.'

Several other seedless white-fleshed clones have arisen throughout the grapefruit producing regions of the world, most notably 'Cecily' from South Africa (8). However, none have surpassed the popularity of 'Marsh,' the first seedless grapefruit known.

'Marsh' became the leading grapefruit cultivar shortly after its introduction. Despite the increasing popularity of pigmented grapefruit, it remains the leading cultivar grown today. The United States leads in world grapefruit production, producing nearly 45% of the world total in the 1990-91 season (4), and Florida grapefruit production accounted for nearly 85% of the U.S. total (15). Over 48% of the Florida grapefruit crop in 1990-91 was 'Marsh.' 'Marsh' likewise predominates in other grapefruit producing countries.

The amount of 'Marsh' grapefruit produced relative to pigmented types has been declining with increased consumer preference for pigmented grapefruit. In 1971, nearly 24,000 hectares (> 52% of the total) of 'Marsh' grapefruit were grown in Florida, compared with less than 12,000 hectares (< 26% of the total) of pigmented grapefruit cultivars, mostly the 'Redblush' or 'Ruby Red' (15). By 1990, 'Marsh' hectarage in Florida decreased to slightly more than 19,000 (> 45% of the total), but pigmented varieties increased to more than 20,500 hectares (nearly 49% of the total).

The decline in overall production and hectarage planted with 'Marsh' will continue. From July 1986 through June 1991, a total of 1,691,272 'Marsh' trees were produced by registered nurseries in Florida (34.8% of all registered grapefruit trees), but 3,087,728 registered pigmented grapefruit trees were grown (63.6% of the total) (Charles O. Youtsey, chief of the Florida Bureau of Citrus Budwood Registration). The pigmented cultivars included 'Redblush' (for 'Ruby Red') 'Star Ruby,' 'Flame,' 'Ray Ruby,' and 'Rio Red.' Although, as a group, the pigmented cultivars are being planted more than 'Marsh,' it is noteworthy that not one of these individually has been planted in greater number than 'Marsh.'

Grapefruit cultivars produce polyembryonic seeds that contain apomictic embryos (of nucellar origin), and few or no zygotic embryos, so most seedling progeny are genetically identical clones of the seed parent tree. 'Duncan' grapefruit has been used as a pollen parent in interspecific *Citrus* hybridization to produce several tangelo cultivars (hybrids of grapefruit with mandarin, *C. reticulata* Blanco). 'Marsh' grapefruit has not been widely used in hybridizations because of poor pollen fertility resulting from either spindle mechanism failure (10) or increased univalency (5). In addition to nucellar embryony and low fertility, those few grapefruit x grapefruit hybrid seedlings that can be produced by sexual hybridization exhibit general characteristics of inbreeding depression, and may not bear typical fruit.

Contemporary scholars of *Citrus* agree that grapefruit is more correctly considered an interspecific hybrid of *C. grandis* (L.) Osb. (pummelo) and *C. sinensis* (L.) Osb. (sweet orange), rather than a "true" species (1, 9, 14). The narrow germplasm base represented among grapefruit cultivars (resulting from its likely interspecific origin and the development of all cultivars from a single germplasm introduction) and the biological factors described above preclude utilization of hybridization and selection among grapefruit for genetic advance and cultivar development. Bowman and Gmitter (2, 3) have documented greater genetic diversity among grapefruit-like *Citrus* clones called "forbidden fruit," "shadette," "wild grapefruit," etc., found growing on various Caribbean islands. Some of these forms may provide breeding parents useful for grapefruit cultivar development, because of their ability to produce zygotic seedlings at frequencies much greater than common grapefruit cultivars (author's unpublished data).

Mutation breeding, in the broadest sense, has been the only method used to develop new grapefruit cultivars. All grapefruit cultivars now

'Marsh' became the leading grapefruit cultivar shortly after its introduction.

> *'Marsh' grapefruit has been the primary germplasm source for the diversification and selection process that has produced nearly all of the pigmented grapefruit cultivars.*

grown originated either as nucellar seedlings or as bud sport mutations that exhibited some desirable phenotypic change; none have come from controlled hybridization. Naturally occurring and induced mutations have been exploited. Growers and scientists have selected new clones primarily on the basis of increased fruit pigmentation.

'Marsh' grapefruit has been the primary germplasm source for the diversification and selection process that has produced nearly all of the pigmented grapefruit cultivars. A 'Marsh' tree in Florida gave rise to a limb sport that bore fruit with slightly-pigmented pink flesh; this limb, discovered in 1913, became the bud source for the 'Thompson' (for 'Pink Marsh') grapefruit (8).

Various limb sports that produced more intensely pigmented fruit were found on 'Thompson' grapefruit trees in Florida and Texas; these included 'Paweett Red,' 'Burgundy,' and 'Redblush' (also known as 'Ruby Red,' the most widely grown pigmented variety) (13). 'Ruby Red' in turn gave rise directly to 'Ray Ruby' via spontaneous bud sport mutation (6) and to 'Rio Red' indirectly via bud sport mutation from an unreleased clone that resulted from irradiation of 'Ruby Red' budwood (7). 'Faweett' gave rise to 'Henderson' via bud sport mutation; 'Flame' was selected from a 'Henderson' nucellar seedling population (13). 'Ray Ruby,' 'Rio Red,' and 'Flame' are the newest and most intensely pigmented cultivars (excluding 'Star Ruby').

With the sole exception of 'Star Ruby' grapefruit, all of the major contemporary pigmented grapefruit cultivars are descended from 'Marsh' by somatic mutation. Although 'Marsh' will decline in significance and production in response to consumer preference for pigmented grapefruit, its influence will continue in the future through the propagation of its somatic offspring.

Literature cited

1) Barrett, H.C. and A.M. Rhodes. 1976. A numerical taxonomic study of affinity relationships in cultivated citrus and its close relatives. *Syst. Bot.* 1:105-136.

2) Bowman, K.D. and F.G. Gmitter, Jr. 1990a. Caribbean forbidden fruit: Grapefruit's missing link with the past and bridge to the future? *Fruit Var. J.* 44:41-44.

3) Bowman, K.D. and F.G. Gmitter, Jr. 1990b. Forbidden fruit (*Citrus* sp., Rutaceae) rediscovered in Saint Lucia. *Econ. Bot.* 44:165-173.

4) Food and Agriculture Organization of the United Nations. 1992. Citrus Fruit, Fresh and Processed, Annual Statistics 1992.

5) Gmitter, F.G., Jr., X.X. Deng, and C.J. Hearn. 1992. Cytogenetic mechanisms underlying reduced fertility and seedlessness in *Citrus. Proc. Int. Soc. Citriculture* (in press).

6) Hensz, R.A. 1978. 'Ray Ruby' grapefruit, a mutant of 'Ruby Red' with redder flesh and peel color. *J. Rio Grande Valley Hort. Soc.* 32:39-41.

7) Hensz, R.A. 1985. 'Rio Red,' a new grapefruit with a deep-red color. *J. Rio Grande Valley Hort. Soc.* 38:75-76.

8) Hodgson, R.W. 1967. Horticultural varieties of citrus. pp. 431-591. *In:* W. Reuther, H.J. Webber, and L.D. Batchelor (eds.), *The Citrus Industry, Vol. I,* Revised Edition, University of California, Riverside.

9) Kumamoto, J., R.W. Scora, H.W. Lawton, and W.A. Clerx. 1987. Mystery of the forbidden fruit: Historical epilogue on the origin of the grapefruit, *Citrus paradisi* (Rutaceae). *Econ. Bot.* 41:97-107.

10) Raghuvanshi, S.S. 1962. Cytogenetical studies in genus Citrus. IV. Evolution in genus *Citrus. Cytologia* 27:172-188.

11) Robinson, T.R. 1933. The origin of the 'Marsh' seedless grapefruit. *J. Hered.* 24:437-439.

12) Robinson, T.R. 1947. Count Odette Phillippi—A correction to Florida's citrus history. *Proc. Fla. State Hort. Soc.* 60:90-92.

13) Saunt, J. 1990. *Citrus Varieties of the World.* Sinclair International Ltd., Norwich.

14) Scora, R.W., J. Kumamoto, R.K. Soost, and E.M. Nauer. 1982. Contribution to the origin of the grapefruit, *Citrus paradisi* (Rutaceae). *Syst. Bot.* 7:170-177.

15) Terry, R.R., C.T. Erick, and E. Leiphart. 1992. Florida Agricultural Statistics, Citrus Summary 1990-1991. Florida Dept. of Agriculture and Consumer Services.

16) Tison, E.H. 1927. *Los Angeles Times.*

17) Webber, H.J. 1928. Notes on the history of citrus varieties. *California Citrograph* 14: 36-39.

Author Gmitter was with the University of Florida-IFAS, Citrus Research and Education Center, Lake Alfred, Florida, when this article originally appeared, July 1993.

The 'Totem' Strawberry

by Hugh Daubeny, F.J. Lawrence, & P.P. Moore

The 'Totem' strawberry cultivar has been a major factor in maintaining the reputation of the Pacific Northwest of North America as a producer of high quality strawberries for the processing market. The cultivar continues to be the most popular throughout the region, which includes southwestern British Columbia, western Washington, and western Oregon. In 1990, 'Totem' accounted for approximately 58% of the certified plant sales throughout the region, with the second and third most widely planted cultivars, 'Sumas' and 'Shuskan,' accounting for approximately 9% and 8%, respectively. The 58% represents more than 20 million 'Totem' plants distributed in that year. In British Columbia, the cultivar has accounted for as much as 80% of the plantings (6). During the past five years, approximately 65% of the harvested strawberry plantings in Oregon have been 'Totem.'

In many respects, 'Totem' is ideally suited to the processing market. It produces relatively firm-textured, conic shape fruit which, at maturity, is a uniform, intense medium to dark red color both internally and externally. This color is considered particularly desirable for many processed products, including yogurt and jam. The firm texture and the uniform color also give an excellent product when sliced and frozen for dessert packs and when used for individual quick freeze (IQF). The texture is such that the integrity of the fruit is maintained upon thawing. The calyx is easily removed at harvest; this is an essential trait for processing. The size is large and is maintained relatively well throughout the season, so that it usually is never less than medium.

This trait, while not absolutely essential for most processed forms, makes for more efficient harvesting.

'Totem' plants have the potential to produce high yields with 5 to 8 tons per acre (11.2 to 17.2 tons per hectare) being common throughout the Pacific Northwest. In Oregon, where the plants have been grown in a modified hill system, at 16-inch (41 cm) spacing in double rows on raised beds, yields of 17.5 tons per acre (39.2 tons per hectare) have been obtained (8).

'Totem' originated from a 1962 cross of two Pacific Northwest cultivars, 'Puget Beauty' and 'Northwest' (1). The seedling was selected in a greenhouse bench test as showing resistance to a composite of races of the red stele causal organism, *Phytophthora fragariae* Hickman (1). The desirable fruit traits of the selection were first seen in 1964. During the subsequent trial period, the selection showed at least as much tolerance to a complex of viruses as one of its parents, 'Northwest,' which at that time was the dominant cultivar in the Pacific Northwest (3). This tolerance, and considerably more winter hardiness compared to 'Northwest,' were important factors in the decision to name the selection (1). Throughout the Pacific Northwest, 'Northwest' was prone to serious winter injury and had a 75% loss in production in two separate years during the 1960s (5).

One reason for the longevity of 'Totem' has been field resistance to red stele (8). However, in recent years, there have been incidences of the breakdown of this resistance. This indicates the occurrence of one or more resistance-breaking races of the causal organism, *P. fragariae*. This is not surprising since, in greenhouse screening tests involving four individual races of the organism, the cultivar was resistant to two of the four, moderately resistant to another, and susceptible to the fourth (2).

'Totem' has resistant reactions to several other diseases including verticillium wilt (*Verticillium albo-atrum* Reinke & Berth.), leaf spot (*Mycosphaerella fragariae* [Tul.] Lindau), and powdery mildew (*Sphaerotheca macularis* [Fr.] Magn.) and an intermediate reaction to leaf scorch (*Diplocarpum earliana* [Ell. & Everh.]) (6). It also shows some resistance to fruit rots (both *Botrytis cinerea* Pers. ex. Fr. and *Rhizopus* spp.) (1, 6). Pre- or at-harvest reaction to *B. cinerea* is at least partly due to fruit scapes, which remain relatively upright until the fruit begins to ripen, and also to the firm fruit texture.

The cultivar is relatively susceptible to two-spotted spider mite, *Tetranychus urticae* Koch (2). This can be a problem with the increasing restrictions on the use of chemicals for control of the pest.

'Totem' is not a prolific runner producer during the year following planting. At times, this has resulted in a shortage of plants available from certified growers. It can also result in relatively low plant numbers in the matted rows, which are traditional in Pacific Northwest strawberry production. This situation can increase the amount of injury after unusually low winter temperatures. The problem of low plant numbers is less likely to occur in subsequent years, and it has often been noted that plantings are more vigorous and

'Totem' originated from a 1962 cross of two Pacific Northwest cultivars, 'Puget Beauty' and 'Northwest.'

> **'Totem' has been used extensively in breeding programs as a source of superior fruit qualities and of virus tolerance.**

uniform in these years. This is also a function of the virus tolerance of the cultivar and can result in higher yields in subsequent years compared to the first.

To ensure good survival, 'Totem' plants from certified growers must be fully dormant when dug for subsequent storage at -1°C (33•F) prior to field planting in the spring (4). This contrasts to the situation with other Pacific Northwest cultivars, including 'Sumas' and 'Shuksan,' plants of which show good survival even if not fully dormant when placed into storage. The fully dormant requirement for 'Totem' can be a problem because conditions are quite frequently unfavorable for digging until growth has commenced in the late winter or early spring.

Another problem with 'Totem' plants can occur because of early flowering. The flowers, which are usually produced above the leaf canopy, are sometimes frost damaged with resulting yield reductions (2). This situation is more common in Oregon than further north.

Although 'Totem' fruit is primarily used for processing, it is acceptable for fresh market, especially if harvested before fully ripe and the calyx is retained. It has a pleasant but not highly aromatic flavor. The relatively concentrated ripening habit, though, is not ideal for the fresh market but certainly is desirable for the processing market, for which large volumes of fruit per harvest make for greater efficiency.

'Totem' has had some success outside the Pacific Northwest. In Britain, it is highly rated by the "pick-your-own" industry for processing as a frozen product. Approximately 250,000 plants of the cultivar are sold annually there (Paul Walpole, personal communication). Success in some other regions has been limited by high temperatures and light intensities, which cause the fruit to darken excessively.

'Totem' has been used extensively in breeding programs as a source of superior fruit qualities and of virus tolerance. It is a parent of the day-neutral cultivar 'Tillikum' from the Washington State University program and of 'Rederest' and 'Bountiful,' both from the Oregon State University, United States Department of Agriculture program. 'Rederest,' which was released in 1989, and now occupies approximately 5% of the plantings in the Pacific Northwest (in 1993), produces fruit suited to processing. It is likely that there will be more cultivars released with 'Totem' or a 'Totem' derivative as a parent.

In 1984, 'Totem' received an Outstanding Cultivar Award from the Canadian Society for Horticultural Science.

'Totem' is likely to remain an important cultivar in the Pacific Northwest for the immediate future. Any replacement cultivar should possess similar fruit qualities with at least as much firmness and less tendency to darken excessively. Plants too should possess qualities similar to those of 'Totem,' with as much or more virus tolerance, but flower later, have the ability for good survival even if not fully dormant when dug for cold storage, have greater runner production in the planting year, and have resistance to more races of the red stele causal organism and resistance or tolerance to two-spotted spider mite.

Literature cited

1) Daubeny, H.A. 1971. Totem strawberry. *Can. J. Plant Sci.* 51:176-177.

2) Daubeny, H.A. 1987. 'Sumas' strawberry. *HortScience* 22:511-513.

3) Daubeny, H.A., R.A. Norton, and B.H. Barritt. 1972. Relative differences in virus tolerance among strawberry cultivars and selections in the Pacific Northwest. *Plant Dis. Reptr.* 56:792-795.

4) Daubeny, H.A. and A.K. Anderson. 1991. Strawberry breeding in British Columbia. pp. 100-101. *In:* A. Dale and J.J. Luby (eds.) *The Strawberry into the 21st Century.* Timber Press, Portland, Oregon.

5) Hancock, J.F., J.L. Mass, C.H. Shanks, P.J. Breen, J.J. Luby. 1991. Strawberries *(Fragaria).* pp. 491-546. In J.N. Moore and J.R. Ballington (eds.) *Genetic Resources of Temperate Fruit and Nut Crops.* Wageningen: ISHS.

6) Lawrence, F.J. 1989. Pacific Northwest strawberry cultivars. *Fruit Var. J.* 43:19-21.

7) Scott, D.H., A.D. Draper, and G.J. Galletta. 1984. Breeding strawberries for red stele resistance. *Plant Breeding Review* 22:195-214.

8) Unger, M. and B.C. Strik. 1988. How I got 17.5 tons per acre of 'Totems.' Lower Mainland Horticultural Improvement Association 30:100-101.

Authors Daubeny, Lawrence, and Moore were with Agriculture Canada Research Station, Vancouver, British Columbia, Canada; National Clonal Germplasm Repository, Corvallis, Oregon; and Puyallup Research and Extension Center, Puyallup, Washington, respectively, when this article originally appeared, October 1993.

The Lingonberry

by Elden J. Stang

The lingonberry (*Vaccinium vitis-idaea* L.) is a woody, evergreen dwarf shrub distributed worldwide in northern temperate, boreal, and subarctic areas. The fruit is an important berry crop, harvested mostly from the wild in Russia, Scandinavia, the Baltic countries, Poland, and, to a lesser extent, in Japan, Germany, Canada, and Alaska (10, 13, 16, 18).

Camp characterized the lingonberry *Vaccinium vitis-idaea* in the *Ericaceae*, subgenus *Vitis-idaea* (Moench) W. Koch (3). Plant stems are semi-woody, bearing numerous shoots one to two millimeters in diameter. Simple, petiolate, evergreen, leathery, obovate leaves alternate in a spiral. Leaf upper surface is dark green; the lower surface is pale green, waxy, with black glandular dots. Plants reproduce by seed and rhizomes from which shoots arise at nodes. Roots consist of tap roots with finely divided rootlets at the extremities and adventitious roots occurring at nodes along creeping stems and rhizomes. Flowers are produced singly or in clusters in terminal racemes, with four locules per ovary, four sepals, a bell-shaped corolla and eight stamens with nonspurred anthers (8). Pollen is borne in tetrads shed through a terminal pore in the anther. The ovary is inferior, producing a true globose berry, carmine in color when ripe, up to 1.2 centimeters in diameter.

Fernald considered the smaller North American form as a variety *V. vitis idaea* L. var. *minus* Lodd. and the larger European plant as the variety *vitis-idaea* (6). Hulten recognized the two races as subspecies with *vitis-idaea* as the larger lowland race and the dwarf arctic montane race as *minus* (Lodd.) Hulten (12). Both are distinguished mainly by plant size. Leaf size in *V. vitis-idaea* may average 2.5 cm in length and 1.0 cm in width, in *V. vitis-idaea* var. *minus* 1.0 cm in length and 0.5 cm in width (28). Plant height for *V. vitis-idaea* may exceed 30 cm; for *vitis-idaea* var. *minus*, height rarely exceeds 20 cm. Both North American and European varieties are reported by various authors to have chromosome numbers $2n = 24$ (8).

Worldwide, at least 25 English names for *V. vitis-idaea* are reported (10). Among the more common are lingonberry, cowberry, moss cranberry, mountain cranberry, red whortleberry, or alpine cranberry. In Newfoundland, the fruit is called the partridgeberry, in Finland puolukka, in Germany the preiselbeere, and in Sweden lingon or lingen. For aesthetic and marketing reasons, Pliszka suggested the name lingonberry be used as the English name rather than cowberry (20).

Numerous uses for lingonberry fruit are reported. In northern Europe and Japan, these include juice, sauce, preserves, candy, jelly, syrup, in ice cream, as pickles, wine, and liqueurs (10, 13, 18, 19). In eastern Europe, extraction of arbutin from the leaves for use as a medicinal for stomach disorders is described (22). Rehder suggested use of the plant as an ornamental ground cover (23).

Despite the fact large quantities of lingonberry fruit are still harvested from the wild, urban encroachment, changes in forest management, variable fruit quality from native stands, and fluctuations in annual yield have stimulated research worldwide on plant improvement and methods for domestication and cultivation of this species (10, 24, 26). The most concerted attempts at domestication were carried out in Finland, Germany, and Sweden, beginning in the late 1960s (7, 14, 15, 16, 17). Subsequent research in North America involved screening of seedlings, characterization of the species and vegetative and reproductive responses (10, 11). Successful small scale commercial production of lingonberries in Germany was first described by Dierking in 1985 (4).

A key to domestication and development of any prospective new crop is a genetic pool of adapted and productive cultivars. Development of lingonberry cultivars has been minimal in comparison to other fruit species and only relatively recently has resulted in a limited series of promising clones. To date, the few cultivars available are exclusively selections from native material. The origin and brief descriptions for nine lingonberry cultivars herein are gleaned from various sources, cited after the description.

'Red Pearl,' 1981. Originator: A. Blanken Boskoop, the Netherlands. Origin not reported. Wide, bushy plants upright in growth habit, height 20 to 30 cm; 5 to 12 fruit in clusters, large fruits (7-12 mm), round ripening in September to October (2).

'Koralle,' 1969. Originator: H. van der Smit, Origin: Reeuwijk, the Netherlands. Origin: not reported. Strongly branched, height up to 30 cm, round fruit in clusters of 5 to 12 berries varying from light red to dark red. Note: 'Koralle' was originally a collection of 35 plants selected from a population of seedlings as a mixture under the name 'Koralle.' Thus, identity may be confused, depending on the clone propagated from the original stock (1).

'Sussi' (BV 401), 1985. Originator: Prof. Sven Dahlbro, Copenhagen. Named and released

Worldwide, at least 25 English names for V. vitis-idaea *are reported, including lingonberry, cowberry, moss cranberry, mountain cranberry, red whortleberry, and alpine cranberry.*

The development of widely adapted lingonberry cultivars for potential use in commercial production of this delightful small fruit is in its infancy.

(patented) by the Swedish University of Agricultural Sciences, Division of Fruit Breeding, Balsgard. Origin: A selection from seed collected in Smaland, Sweden. Lowbush plant form with erect shoots 15 to 25 cm height, flowering late May in southern Sweden, berries globular, dark red, large (0.4 g/fruit), uniformly ripening about August 20 in Sweden. Productive, averaging 11 fruit per cluster (27).

'Sanna' (BV 35), 1987. Originator: Prof. Sven Dahlbro, Copenhagen. Origin: A selection from seed collected in Smaland, Sweden. Named and released (patented) by the Swedish University of Agricultural Sciences, Balsgard. Erect shoots 15 to 25 cm in height, flowering late May in southern Sweden. Fruits red, globose, ripening in mid-August. Fruits comparable in size to 'Sussi' (5).

'Scarlet,' date and originator unknown. Origin: Norway. A selection from seed suggested for use as a pollinator for 'Koralle.' Plant height 30 to 38 cm, spread 38 to 45 cm at maturity (9).

'Erntedank,' 1975. Originator: Albert Zillmer, Uchte, Germany. Origin: Plant selected from the wild in a upland moor west of Uchte, Germany. Moderate growth, fruit small to medium sized, very productive, producing both a spring and summer crop (29).

'Erntekrone,' 1978. Originator: Albert Zillmer, Uchte, Germany. Origin: Selected from the wild near Uchte, Germany. Vigorous, stiff and rigid upright growth, leaves somewhat more circular compared to characteristic oval shape in lingonberry. Fruit large, dark red, highly productive. Summer crop not as consistent as for 'Erntedank' (29).

'Erntesegen,' 1981. Originator: Albert Zillmer, Uchte, Germany. Origin: Selected from the wild near Uchte, Germany. Long, soft shoots with unusually large leaves. very large fruit, some exceeding 1.0 cm in diameter, bright red, productive, suggested by the originator for commercial production, patented (29).

'Masovia,' 1985. Originator: Lech Kawecki, Poland. Origin: Selected from the wild in the Lasy Bolimowskie forest, 60 km west of Warsaw in 1981. Vigorous growth and rhizome production, heavy fruit set. Named and released by K. Pliszka, Warsaw Agricultural University (21).

Since 1984, our preliminary screening trials and research with lingonberry in Wisconsin have suggested the potential for cultivation in the northern United States (25). Sandy acidic soils in central Wisconsin with ample water for irrigation provide virtually unlimited suitable sites for cultivation of the plant.

Current objectives in our program are to screen cultivars, seedling, and clonally propagated lingonberry populations for adaptability and productivity; to evaluate critical cultivation practices including soil amelioration with organic matter, mulches, herbicides for weed control, and to determine plant nutrition and irrigation water requirements. Establishment of small-scale grower demonstration trials concurrent with studies of fruit processing requirements and market development are subsequent objectives, as recently described by St. Pierre for native fruit species in Saskatchewan (26).

In 1987, I had the opportunity under the auspices of a Fulbright research grant to spend a four-month sabbatical at the Agricultural Research Centre, Department of Horticulture, Pinkkio, in southwest Finland in collaboration with Professors Jaakko Säkö, Heimo Hiirsalmi, and others. My primary objective was to collect and catalogue native Finnish lingonberry plant material for further testing in North America.

Characteristics such as plant vigor, freedom from obvious pests, flowering period, approximate ripening date, number of fruit per cluster, and general fruit size were recorded for 122 individual plant selections and 22 collections of seed from individual plant clones or pooled seed sources.

From the seed collections, approximately 15,000 seedling lingonberry plants were established at the Hancock Experiment Station, Hancock, Wisconsin, in 1988. Since then, 17 selections have been made from this planting, most for inclusion in advanced screening trials. Of the initial whole plant selections, six outstanding clones were chosen to be included in a screening trial.

Two of these, WI 102 and WI 108, have displayed hardiness, precocity, excellent fruit size and fruit color. Initial yields of WI 102 and 108, albeit limited, were comparable to 'Erntedank.' Naming and release of these clones by the University of Wisconsin, Madison, is proposed for 1993. Of the named cultivars in the earliest selection trial terminated in 1991, 'Erntedank' appears to be better adapted, more hardy and productive than 'Koralle,' 'Erntesegen,' or 'Erntekrone.' 'Sussi,' 'Sanna,' and 'Masovia' have not as yet fruited sufficiently for a meaningful comparison.

The development of widely adapted lingonberry cultivars for potential use in commercial production of this delightful small fruit is in its infancy. Substantial further testing of the limited number of cultivars currently available is necessary. Regardless, our early results with the few cultivars available and our advanced selections suggest great promise for further development of *Vaccinium vitis-idaea* as a new crop for cool temperate regions where blueberries and cranberries already are important in berry production.

Literature cited

1) Anon. 1970. *Vaccinium vitis-idaea* 'Koralle' (Dutch). Dendroflora 7:85-88.

2) Anon. 1982. *Vacciniun vitis-idaea* 'Red Pearl' (Dutch). Dendroflora 19:90.

3) Camp, W.H. 1945. The North American blueberries with notes on other groups of *Vacciniaceae*. Brittonia 5:203-275.

4) Dierking, W. 1985. Ten years experience in lingonberry production. *Acta Hort.* 165:269-271.

5) Eckerbom, C. 1987. 'Sanna.' BV35—a new lingonberry from Balsgård. Sveriges Lantbruksuniversitet. Växtförädling av Frukt och Bär, Balsgård. Verksamhetsberättelse 1984-1985. pp. 133-134 (English summary).

6) Fernald, M.L. 1970. *Gray's Manual of Botany*. 8th ed. D. Van Nostrand Co., New York.

7) Ferngvist, I. 1977. Results of experiments with cowberries and blueberries in Sweden. *Acta Hort.* 61:295-300.

8) Hall, I.V. and J.M. Shay. 1981. The biological flora of Canada. 3. *Vaccinium vitis-idaea* L. var. *minus* Lodd. supplementary account. *Can. Field-Nat.* 95:434-464.

9) Hartmann's Plantation, Inc., Grand Junction, Michigan.

10) Holloway, P.F. 1981. Studies on vegetative and reproductive growth of lingonberry (*Vaccinium vitis-idaea* L.). Ph.D. Diss., Univ. of Minnesota, St. Paul.

11) Holloway, P.F., R.M. Veldhuizen, C. Stushnoff, and D.K. Wildung. 1982. Vegetative growth and nutrient levels of lingonberries grown in four Alaskan substrates. *Can. J. Pl. Sci.* 62:969-977.

12) Hulten, E. 1970. The Circumpolar Plants II. Dicotyledons. Almqvist and Wiksell, Stockholm.

13) Iwagaki, H., S. Ishikawa, T. Tamada, and H. Koike. 1977. The present status of blueberry work and wild *Vaccinium* spp. in Japan. *Acta. Hort.* 61:331-334.

14) Lehmushovi, A. 1977. Trials with the cowberry in Finland. *Acta Hort.* 61:301-308.

15) Lehmushovi, A. and H. Hiirsalmi. 1973. Cultivation experiments with the cowberry. Significance of substrate, liming, fertilization, and shade. *Ann. Agric. Fenn.* 12:95-101.

16) Lehmushovi, A. and J. Säkö. 1975. Domestication of the cowberry (*Vaccinium vitis-idaea* L.) in Finland. *Ann. Agric. Fenn.* 14:227-230.

17) Liebster, G. 1975. Growing red whortleberries (*Vaccinium vitis-idaea*) on cultivated land—a new objective of experimental research work in fruit growing. *Erwerbsobtbau.* 17:39-42, 58-61.

18) Liebster, G. 1977. Experimental and research work on fruit species of the genus *Vaccinium* in Germany. *Acta Hort.* 61:19-24.

19) Müller, H.P. 1977. The use of *Vaccinium* fruit in the dairy industry. *Acta Hort.* 61: 343-347.

20) Pliszka, K. 1985. Foreword, Third lnt. Symp. *Vaccinium* culture. Warsaw, Poland. *Acta. Hort.* 165:7.

21) Pliszka, Kaziemierz, and Lech Kawecki. 1985. 'Masovia'—a new Polish selection of lingonberries. *Acta Hort.* 165:273.

22) Racz, G., I. Fuzi, and L. Fulop. 1962. A method of determination of the arbutin content of cowberry leaves (*Folium vitis-idaea*). *Rumanian Med. Rev.* 6(1):88-90 (Abstr.).

23) Rehder, A. 1940. *Manual of Cultivated Trees and Shrubs.* Macmillan, New York.

24) Scibisz, K. and K. Pliszka. 1985. Effect of mulching and nitrogen fertilization upon growth and yield of lingonberries (*Vaccinium vitis-idaea* L.). *Acta Hort.* 165:275-277.

25) Stang, E.J., G.G. Weis, and J. Klueh. 1988. Lingonberry: potential new fruit for the northern United States. In: J. Janick and J.E. Simon, eds. *Advances in New Crops. New Crops: Research, Development, Economics.* Timber Press, Portland, Oregon pp. 321-323.

26) St. Pierre, Richard G. 1992. The development of native fruit species as horticultural crops in Saskatchewan. *HortScience* 27(8): 866, 947.

27) Trajkovski, V. 1985. 'Sussi' (BV 40, the first cultivar of lingonberry from Balsgård). Sveriges Lantbruksuniversitet. Växtförädling av Frukt och Bär, Balsgård. Verksamhetsberättelse 1984-1985, pp. 132-134 (English summary).

28) Welsh, S.L. 1974. *Anderson's Flora of Alaska and Adjacent Parts of Canada.* Brigham Young Univ. Press. Provo, UT.

29) Zillmer, Albert. 1985. Account of my three types of *Vaccinium vitis-idaea*, 'Erntedank,' 'Erntekrone,' 'Erntesegen.' *Acta Hort.* 165:295-297.

Author Stang was with the Department of Horticulture, University of Wisconsin, Madison, when this article originally appeared, January 1994.

The 'Granny Smith' Apple

by Ian J. Warrington

'Granny Smith,' like many of the current commercial apple cultivars, arose as a chance seedling. It has been in cultivation for over 120 years but did not achieve significant volumes on international markets until after 1950. It is grown widely in many countries throughout the world but is particularly important in the Southern Hemisphere (i.e., Australia, Argentina, Chile, New Zealand, and South Africa) where production in 1991 was estimated to exceed 50 million cartons (one carton = approximately 18.5 kg). It currently accounts for about 7% of United States production and about 4% of European apple production.

Origins

There have been many interesting and different stories about the origins of the cultivar, but the most authentic and plausible appears to be that told by Mr. Benjamin Spurway, a grandson of Maria Ann "Granny" Smith, after whom the cultivar is named (13). "Granny" Smith was born Maria Ann Sherwood in 1800. She married Thomas Smith in Sussex, England, and emigrated to Australia in 1838. With a young family of six children, they settled in the Ryde district near Sydney, and later acquired a small block of land in Eastwood, which they established as an orchard and market garden. Mrs. Smith assumed the weekly task of making the arduous 15-mile trip by horse-drawn cart to the Sydney markets to sell the family's produce. On one such trip, she was given some apples, apparently of the cultivar 'French Crab' from Tasmania, by a fellow stall holder (Mr. Thomas Lawless) to test their cooking quality. On returning home, she baked a pie from this fruit but deliberately sowed the seeds in the cultivated ground outside her kitchen window. Only one seed germinated and it was left undisturbed to grow. When the tree bore fruit, Mrs. Smith realized that it produced very fine apples that both cooked and stored well, and consequently, she nurtured the tree and harvested the fruit for her own use. Mr. Spurway proudly noted that "it is the best cooking apple, the best carrier, best drying apple, and if allowed to remain on the tree until thoroughly ripe, then picked and placed in straw until it mellows, it takes its place amongst the best eating" (13).

An alternative version to this story is that told by a Mr. T. Small, who claimed that, as a 12-year-old boy in 1868, he had gone with his fruit-growing father Mr. E.H. Small, to visit the Smiths. Mrs. Smith had taken them to see the apples growing on a tree on the property. This tree had apparently grown from some rotting 'French Crab' apples in gin cases that Mrs. Smith had brought back from the market and discarded near the creek on her property. One seedling survived and grew, and this eventually became the 'Granny Smith' known today (8, 9). By the 1860s, Mrs. Smith had long been a popular figure in the Ryde-Eastwood district and now an elderly lady, she was affectionately known as "Granny." She died in 1870.

'Granny Smith' is, therefore, widely regarded as being the progeny of open-pollinated 'French Crab,' but the true parentage cannot be confirmed. Some of the American 'Greening' apple cultivars have similar features. 'Cleopatra' is also very similar and may have come from a common source to 'Granny Smith.'

Mrs. Smith did not exploit the cultivar but gave wood for grafting or budding to those who asked for it. It was, in fact, Mrs. Smith's sons-in-law James Spurway and Henry Johnston, and another local orchardist, Edward Gallard, that began the first significant cultivation of the cultivar. It was first recorded by the Royal Horticultural Society in 1883 (12). In 1895, a *New South Wales Agricultural Gazette* article on export fruit by Albert H. Benson, then fruit expert of the Department of Agriculture, lists 'Granny Smith Seedling' as a late cooking apple suitable for export (3, 9). Benson was apparently the first to arrange for the cultivar to be grown outside of the district where it had originated. A color plate of 'Granny Smith' was published in the *N.S.W. Agricultural Gazette* in September, 1904 (opposite page 910). The commercial possibilities for the cultivar were recognized by nurserymen at about the turn of the century, and significant plantings occurred from the 1920s onward. By 1975, 39% of Australian apple production was from 'Granny Smith,' and that proportion remained in 1990-91 (9).

The commercialization of 'Granny Smith' also occurred rapidly in New Zealand. The Auckland nurserymen Mr. David Hay & Son of Montpellier Nurseries, were advertising 'Grannie Smith' for sale in 1895 and in their 1905-06 general catalogue, described the cultivar as:

"A very large, green, handsome, long-keeping cooking apple. Superior to French Crab. The fruits resemble Cleopatra in shape and general appearance, but are much larger. A valuable addition to late keepers."

At that time, trees were sold at the following prices: "Price, 1s [approx 5¢ US] each; 10s to 12s [50¢ to 60¢ US] per dozen, unless where otherwise quoted. Large purchasers liberally dealt with."

Their catalogue also stated boldly: "Winter's coming on! If the cellar is full of apples, the children won't go short of pies."

Another important nurseryman from New Zealand, Mr. Hayward Wright (responsible for the selection of 'Hayward' kiwifruit), actively promoted the use of 'Granny Smith' in the 1920s. It was approved for export from New Zealand to Europe in 1923, and rapidly became an important commercial cultivar. Throughout the last decade, it has been the most important single cultivar, accounting for about 40% of total New Zealand apple production. Major expansion of 'Granny Smith' in North America and Europe occurred much later in the 1970s and was largely the result of demand generated through the popularity of the cultivar among consumers of fruit exported from the Southern Hemisphere. It is currently (in 1994) the third largest volume apple cultivar shipped from Washington State (6).

Plantings worldwide of 'Granny Smith' are predominantly of the standard nonspur type. Spur types such as 'Greenspur' and 'Granspur' have been selected and patented (1, 15), but these have a reputation for producing lower yields and poorer quality fruit than the standard type.

'Granny Smith' appears to have been used as a parent in some early apple breeding programs. A cultivar named 'Granny Mack' was listed in nursery catalogues in New Zealand in the 1940s, and was described as a cross between 'Granny Smith' and 'McIntosh'—the fruit of this late dessert cultivar was apparently similar in appearance to 'Granny Smith' but had a red rather than green skin color (16).

Fruit description

Fruit are round conical in shape, fairly regular, somewhat flattened at the base and apex, and slightly five-crowned at the apex (4). Fruit size is classed as medium, and, under commercial production in New Zealand, the median size is in the 100 to 113 fruit/carton count (165-205 g) (14).

The fruit skin has a bright green ground color that becomes greenish yellow with maturity. There are no stripes, and lenticels can be very conspicuous as numerous large areolar dots. The skin is smooth and quite greasy in advanced maturity. The flesh is greenish white, firm, and rather coarse-textured. It is a juicy, subacid, and refreshing apple, but lacks flavor (4, 12).

Growth habit and yields

'Granny Smith' has been classified as a Type IV cultivar with an acrotonic growth tendency (7), i.e., in a natural growth form, trees develop as an inverted cone. The majority of fruit buds are borne on one- and two-year-old wood, initially in terminal positions, but later in lateral positions on horizontal and inclined (bending) branches. Spurs develop easily, although significant zones of "blind" wood can persist. With such a growth habit, heading cuts will stimulate strong

Mrs. Smith did not exploit the cultivar but gave wood for grafting or budding to those who asked for it.

> *Premature harvest, especially of solidly green-colored fruit, will result in poor eating quality, reduced storage longevity, and enhanced storage scald development.*

vegetative growth rather than encourage fruit bud formation (8).

In many New Zealand fruit growing districts (such as Hawke's Bay), growth is moderately vigorous on rootstocks such as MM.106, but excessively vigorous on rootstocks such as Northern Spy, M.12, and M.16.

'Granny Smith' is partially self-sterile, and compatible pollen from other diploid cultivars such as 'Red Delicious,' 'Rome Beauty,' 'Akane,' 'Golden Delicious,' 'Criterion,' and 'Winter Banana' must be provided through an adequate interplanting arrangement (1). In most years, 'Granny Smith' is self-thinning, but it can have a tendency toward biennial bearing. Bloom thinning with chemicals is practiced but must be carried out with caution as overthinning can easily result from inappropriate timing or application.

Yields of 'Granny Smith' of 2,500 to 3,000 bushels/acre (120-130 tonnes/ha) can be routinely achieved under commercial production systems in New Zealand on orchards which are more than 10 years old (10, 14). Lower yields are achieved in both North America and Europe. The record crop for a mature commercial 'Granny Smith' orchard in New Zealand is 4,000 bushels/acre (10).

Maturity

'Granny Smith' has long been regarded as an excellent multipurpose apple, having both good dessert (fresh) and culinary properties.

The cultivar is classed as being late to very late (12). Fruit is harvested in New Zealand in April-May, and in the Pacific Northwest, it is regarded as a 180-day apple (i.e., days from full bloom to harvest), resulting in an October to November harvest (1, 6). Harvest is usually carried out with a once-over pick.

Ethylene production cannot be used as a maturity marker for 'Granny Smith' (unlike some other cultivars such as 'Cox's Orange Pippin' and 'Braeburn'), as it remains very low throughout the commercial harvest period (15).

Premium postharvest storage quality is achieved by harvesting fruit with a medium (not dark) green skin color and a starch-iodine index number of one or higher for 95% of the fruit; fruit should not be harvested until starch has begun to disappear from the core (2, 6, 11). Premature harvest, especially of solidly green, colored fruit, will result in poor eating quality, reduced storage longevity, and enhanced storage scald development.

Acknowledgments

The following are thanked for their assistance in the preparation of this article: A.R. (Ross) Ferguson, Richard J. Mangin, Alan E. Scott, R. Sweedman, W.K. (Bill) Thompson, Jane Turner, and W.J.W. (John) Wilton.

Literature cited

1) Ballard, J.K. 1981. 'Granny Smith.' An important apple for the Pacific Northwest. Ext. Bul. 0814, Wash. State Univ.

2) Beattie, B.B., B.L. Wild, and G.G. Coote. 1972. Maturity and acceptability of early-picked 'Granny Smith' apples for export. *Australian J. Exp. Agric. and An. Husb.* 12:323-327.

3) Benson, A.H. 1895. Fruits to export, and how to export them. *Agricultural Gazette of N.S.W.* 6:554-566.

4) Bultitude, J. 1983. 'Granny Smith.' p. 180. In: *Apples. A guide to the identification of international varieties.* The Univ. of Washington Press, Seattle.

5) Kilpatrick, D.T. 1964. New semi-dwarf "spur type" 'Granny Smith.' *South Australian J. Agric.* 68(2):57, 59-61.

6) Kupferman, E.M. 1992. Maturity and storage of apple varieties new to Washington State—1992. *Washington State Univ. Tree Fruit Postharvest J.* 3(1:9-16.

7) Lespinasse, J.M., P. Chol, J. Dupin, and E. Terrenne. 1977. La conduite du Pommier. Types de fructification, incidence sur la conduite de l'arbre. Institut National de Vulgarisation pour les Fruits, Legumes et Champignons, Paris.

8) Letters, R. 1962. The 'Granny Smith' apple in Tasmania. *Tasmanian J. Agric.,* May: 185-191.

9) McAlpin, M. 1976. 'Granny Smith'—Australia's most famous apple. *The Fruit World and Market Grower,* June: 12.

10) McKenzie, D.W. 1985. Training essential for early Granny production. *The Goodfruit Grower,* Jan.: 8, 10-11.

11) Reid, M.S., C.A.S. Padfield, C.B. Watkins, and J.E. Harman. 1982. Starch iodine pattern as a maturity index for 'Granny Smith' apples. I. Comparison with flesh firmness and soluble solids content. *New Zealand J. Agric. Res.* 25: 239-243.

12) Smith, M.W.G. 1971. 'Granny Smith.' p. 222. In: *National Apple Register of the United Kingdom.* Ministry of Agric., Fisheries and Food, London.

13) Spurway, B. 1967. The 'Granny Smith' apple. Transcript of talk to Eastwood Senior Citizens Club (unpublished; File CR/l/10 HortResearch, Palmerston North, New Zealand).

14) Warrington, I.J, C.J. Stanley, D.S. Tustin, P.M. Hirst, and W.M. Cashmore. 1989. Influence of training system on 'Granny Smith' yield and fruit quality. *Compact Fruit Tree* 22: 12-20.

15) Watkins, C.B., J.B. Bowen, and V.J. Walker. 1989. Assessment of ethylene production by apple cultivars in relation to commercial harvest dates. *New Zealand J. Crop and Hort. Sci.* 17: 327-331.

16) Wright, H.R. 1942. General catalogue of fruit, shelter and ornamental trees; hedgeplants, roses, etc.; strawberries, gooseberries, blackberries, loganberries, raspberries, currants, rhubarb. Auckland, Avondale Nurseries.

Author Warrington was with the Horticulture and Food Research Institute of New Zealand, Ltd., Batchelar Research Centre, Palmerston North, New Zealand, when this article originally appeared, April 1994.

The 'Reyna' ('Alfajayucan') Cactus Pear

by Candelario Mondragón Jacobo & Salvador Pérez González

National cactus pear (*Opuntia amyclaea* Tenore) consumption is dominated by the white-fleshed variety 'Reyna' or 'Alfajayucan,' which is an alternative crop for regions regularly affected by drought, where chronic losses in traditional crops such as maize, beans, wheat, and barley are serious economic hardship to farmers. Its wide range of adaptation, stability, and outstanding fruit quality expressed as size, thin peel, reduced seed number, juiciness, and sweetness, has made it the most popular variety among growers and consumers. However, its relatively thin peel demands special management practices after harvest. Even though 'Reyna' is susceptible to *Alternaria* spp. ('Fiebre del oro') and to low winter temperatures, it accounts for 36% of the cactus pear production in Mexico (3).

Cactus pears are naturally distributed in the semiarid regions of Central and Northern Mexico, and they were cultivated well before the discovery of America (1). During the last four decades, cultivation has increased tenfold and is gaining great importance as an alternative crop for the semiarid regions. However, no formal descriptions of cultivated clones have been reported. This work describes the most important cactus pear cultivar to establish a reference for future cultivar descriptions.

Origin and distribution

'Reyna' was selected by farmers before the 1950s from a wild population because of its yield potential and fruit size, thin peel, reduced number of seeds, juiciness, and sweetness. Initially, it was planted in small numbers, but during the last two decades, it has acquired commercial importance.

Current distribution of this cultivar includes approximately 15,000 hectares (37,500 acres) in five different states in Central Mexico. It is a semiarid region located at the southeast tip of the Sonoran-Chihuahuan Desert, at an elevation ranging from 1,800 to 2,400 and a temperate climate and an annual rainfall from 400 to 700 mm. The region was formerly devoted to basic crops and grazing, but it is marginal because soils are shallow, slightly alkaline to acid, and terrain is sometimes steep.

Description

Plants of 'Reyna' are vigorous, reaching 2.5 meters in height and a canopy of 4 to 5 meters in diameter. Cladodes are oval, with a green-blue color, and have three to four persistent spines at the base of each areole.

Flowers are deep orange, and bud break generally starts in January, extending to mid-March, as influenced by weather conditions during winter. If first blossoms are destroyed by frosts, a second flush of floral buds are developed to bear a normal crop (3). However, because 'Reyna' blooms early, it is not cultivated in the northern part of the cactus pear area of distribution.

'Reyna' is an early ripening cultivar with a wide blossom and harvest period (from late June to early September) and a peak in July. Yields are high (10 tons/ha) for a semiarid ecosystem under conventional management practices. They can reach up to 40 tons/ha if supplemental irrigation and application of organic matter are supplied (2).

Fruit quality is high in 'Reyna' (Tables 1 and 2), as determined by several traits such as size and its relatively thin peel that is highly appreciated in the national market. However, even though internal fruit quality is not affected, its thin peel allows bruising and represents a problem for handling and packing.

Fruits are oval flat with green yellow peel and a bright green sweet flesh. The edible portion includes the seeds, which in 'Reyna' are fewer and smaller (16% of aborted seeds).

During the 1992 season, several traits related with fruit quality were observed in 13 different orchards cultivated with 'Reyna' in two of the main cactus pear producing states in Central Mexico (Tables 1 and 2). Sample size was 40 fruits per orchard. Sites 1 to 6 are small orchards (less than 3 hectares) in Las Piramides, near Mexico City, at a higher altitude (2,200 to 2,400 m) and a cool environment (mean annual temperatures between 14 and 18°C [57 and 64°F]). Orchards are generally well fertilized, supplied with manure and yield more than 10 ton/ha. Orchards 7 to 13 are larger (more than 10 ha.), located in poorer soils in a warmer region (mean 18 to 20°C [64 to 68°F]) in the northwest (Guanajuato) at a lower altitude (1,800 to 2,100 m). These orchards have lower yields (3 tons/ha). Sample size was 40 fruits per orchard.

Stability, as determined by the values of coefficients of variability (CV) within and among orchards, was high (low CV) for all traits observed, particularly the flesh to fruit weight

TABLE 1.

Average quantitative traits of ripe fruit of prickly pear, cv. 'Reyena,' registered in 13 orchards in Central Mexico, 1992.

Variable	Mean values	CV (S)
Weight	141.1	17.5
Size		
width	8.6 cm	8.5
length	5.8 cm	8.1
Edible portion	67.0%	17.7
Total seed weight	5.2 g	16.0
Proportion of aborted seeds	16.0%	35.7
Peel thickness	0.47 cm	5.2
Soluble solids	14.8°Brix	5.8

TABLE 2.

Average values for five fruit quality traits in prickly pear cv. 'Reyna' as registered in 13 orchards cultivated in Central Mexico, 1992.

Orchard	Fruit weight g	CV' %	Flesh weight g	CV %	Flesh to fruit %	Seed to flesh %	Soluble solids °Brix	CV %
1	160.4	21.0	103.0	27.3	64.4	5.3	15.0	9.3
2	150.1	7.8	96.8	12.7	64.6	5.0	14.8	8.2
3	122.1	11.8	77.2	17.1	63.3	6.9	14.1	9.3
4	133.0	9.2	82.8	11.9	62.4	5.6	15.6	4.6
5	146.0	15.0	92.5	15.0	63.3	5.9	14.3	6.5
6	145.5	4.2	97.9	12.4	67.6	5.9	15.1	5.6
7	147.1	12.3	99.8	18.0	67.3	5.7	14.4	6.5
8	161.7	19.9	109.7	33.4	67.9	4.4	16.2	3.5
9	150.7	15.1	92.2	16.7	62.0	5.3	14.9	1.0
10	122.7	11.4	71.6	12.8	58.4	6.4	14.0	4.5
11	145.5	12.7	94.6	14.4	65.0	5.7	14.9	1.8
12	117.0	23.6	79.3	27.8	67.5	4.7	15.2	4.5
13	132.9	11.0	79.9	12.6	60.1	6.4	14.5	5.0
Mean	141.1		90.5		64.2	5.6	14.8	
CV	10.2		12.6		4.6	12.6	4.2	

1) *Coefficient of variability between fruits in an orchard.*
2) *Coefficient of variability between orchards.*

Authors Jacobo and González were with the Nopal and Fruit Breeding Program, respectively, Campo Experimental del Norte de Guanajuanto (CENGUA)-INIFAP, San José Iturbide, Gto. Mexico, CP, when this article originally appeared, July 1994.

ratio and soluble solids (less than 10% CV). This explains why 'Reyna' has become so popular among growers in different production systems.

The proportion of aborted (smaller) seeds per fruit is a desirable commercial trait, however, it is strongly influenced by the environment, with a CV of 35%. Other traits with intermediate variability across orchards (CV between 10 and 20%) are fruit weight and total seed weight, which are probably influenced by rain, fruit load per plant, and management practices.

Because more than 80% of the national market is dominated by white-fleshed fruit and the rest by yellow, orange, and red types, 'Reyna' will continue to lead in the prickly pear market until new varieties with thicker peel, and higher resistance to frosts and *Alternaria* spp. are generated through breeding programs.

Literature cited

1) Bravo, H.H. 1978. Las cactáceas de México. Vol 1. Universidad Autónoma de México. México, D.F., México.

2) Mondragón, J.C. 1992. Variedades comerciales y semicomerciales de nopal tunero en el Norte de Guanajuato. Quinto Congreso Nacional y Tercero Internacional sobre Conocimiento y Aprovechamiento del Nopal. Universidad Autónoma Chapingo, 11 al 15 de Marzo de 1992, Chapingo, Texcoco, México.

3) Pimienta, B.E. 1990. El nopal tunero. Colección Tiempos de Ciencia. Universidad de Guadalajara, Guadalajara, Jal., México. SARH, 1992, Sistema producto nopal tuna. Datos basicos. Dir. Gral de Politica Agricola. SARH. Documento de Circulación Interna. Mexico, D.F., México.

The 'Schley' Pecan

by Darrell Sparks

Until this cultivar became susceptible to scab, 'Schley' was regarded by the industry as the standard of quality and, by far, was the cultivar most preferred by the consumer (20). As a result, 'Schley' was widely planted. 'Schley's' popularity decreased greatly after becoming susceptible to scab. Many trees were cut down or else top-worked to less susceptible cultivars. However, the advent of effective fungicides and improved spray machines in the early 1970s brought about renewed interest in 'Schley' as a commercially important cultivar. Once scab was controlled, growers found that existing plantings of 'Schley' were profitable. Nevertheless, 'Schley' is not generally planted in new orchards. The major reason for the lack of many new plantings is the long and well-established record of scab susceptibility of this cultivar. In the early 1900s, 'Schley' was widely planted in the southeastern United States and especially in Georgia. In old orchards in Georgia, 'Schley' is second only to 'Stuart.'

According to Taylor (24), 'Schley' is a seedling grown from an open-pollinated nut of 'Stuart.' This is difficult to believe because 'Schley' has no obvious 'Stuart' characteristics. Regardless, the tree grew from a nut planted in about 1881 by A.G. Delmas, Scranton, Jackson County, Mississippi. The original tree produced 125 pounds of nuts when it was 25 years old.

In 1898, Delmas named the tree 'Schley' in honor of Admiral Winfield Scott Schley, commander of the U.S. Naval Forces during the Spanish-American War. Delmas began to propagate 'Schley' by top-working in 1900.

D.L. Pierson, Monticello, Florida, obtained scions from the original tree in 1902. Pierson intro-

FIGURE 1.
Characteristically scaly bark of a mature 'Schley' trunk. Sprouts often develop from latent buds; sprouting from the trunk is characteristic.

FIGURE 2.
Tree form of mature 'Schley.' Note the arched growth pattern of the lower scaffold limbs. Profuse branching contributes to the closed canopy.

duced the cultivar as 'Admiral Schley,' and the cultivar was originally disseminated under this name (14, 24). 'Schley' is one of the best-known cultivars and is widely disseminated, especially in the southeastern United States. 'Schley' has been used extensively in breeding, and six cultivars with 'Schley' parentage have been released ('Apache,' 'Cherokee,' 'Oconee,' 'Shawnee,' 'Sioux,' and 'Woodroof'). In addition, 'Cape Fear' is a seedling of 'Schley' (7), 'Mahan' is possibly a selfed 'Schley' (26), and, most probably, 'Schley' is a parent of 'Forkert' (5), 'Moreland' (19), and 'Harris Super' (3).

Bud break of 'Schley' begins about four days before that of 'Stuart' (21). Consequently, 'Schley' is more susceptible to late spring freezes than 'Stuart' (27). The foliage is glossy, but tends to be yellowish as in 'Cape Fear' (4). Under conditions of prolonged periods of high light intensity, the leaves of 'Schley' may become somewhat bleached in appearance. Sometimes, this symptom is mistakenly diagnosed as potassium deficiency. Leaf retention in the fall is average (15). Tree vigor is less than 'Stuart' (1). The mature bark of 'Schley' is very scaly (Fig. 1).

The tree is a dense (4), vigorous, and very spreading grower. The lower scaffold limbs on mature trees grow in an arched pattern (Fig. 2). This pattern arises from the fact that as a young tree the growth habit is upright, but, as the tree gets larger, the branches bend under their own weight. Also, in mature trees, the fine branched, fanlike limbs often overlap each other with the result that denseness of canopy is increased and shading out often occurs. Shading out becomes noticeable beginning in July-August (Georgia). Leaves on severely shaded limbs turn bright yellow, defoliate prematurely, and the limbs eventually die. Limb death of this type sometimes causes alarm to the grower; however, the shading out and death of limbs is an inherent characteristic of 'Schley.' Because of the dense spreading canopy, the highest yielding 'Schley' orchards are those that have adequate space between trees to allow the canopy to receive near maximum exposure to sunlight. The dense canopy also makes scab control more difficult than in an open canopy.

The dense canopy, resulting from overlapping limbs, should theoretically make this cultivar a good candidate for selective limb removal. However, experience has shown that this is not the case, as the space created by limb removal is quickly filled in by growth from latent buds along the exposed limb. The tendency for 'Schley' to break latent buds is also evident when mature trees are cut down during orchard thinning. Sprouts develop more readily from 'Schley' stumps than from most other cultivars. Also, sprouting from the trunk is common (Fig. 1). This tendency to sprout has been used to an advantage in attempts to propagate pecans by tissue culture (32). 'Schley' tissue responds more readily than that of some other cultivars.

The dense and spreading canopy makes 'Schley' susceptible to blowing over during severe windstorms. This was graphically demonstrated in windstorms that hit mixed plantings of 'Stuart' and 'Schley.' Tree loss in Georgia was much greater in 'Schley' than in 'Stuart' (22). Similarly, tree loss was greater in 'Schley' when Hurricane Hugo hit a mixed 'Stuart'-'Schley' orchard in South Carolina.

'Schley' is protogynous. 'Schley' is a good late-season pollinizer in the Las Cruces, New Mexico area (18), but is less effective in other areas (13, 17, 31). In Georgia, 'Schley' is a good pollinizer for 'Stuart.' 'Schley, in turn, is pollinated by either 'Cape Fear' or 'Desirable.' (31).

'Schley' precocity is similar to 'Stuart,' but yielding capacity as a mature tree is about 20% less than 'Stuart' (21). The low-yielding capacity is a disadvantage of 'Schley.' However, the yield data for 'Schley' vs. 'Stuart' may be misleading, because the yield comparisons were mostly made prior to the use of effective sprayers and fungicides. That is, 'Stuart' yields may have been higher because of having resistance to scab, whereas 'Schley' did not. Observations in well-managed orchards repeatedly show that 'Schley' is a more consistent producer than 'Stuart, which may increase its long-term yield average.

As with 'Stuart,' the importance of shoot length cannot be overemphasized. This is because pistillate flower formation and fruit cluster size increase greatly with shoot length (Fig. 3). Thus, vigorous shoot growth must be maintained to obtain high yields and to minimize alternate bearing.

The fruit is elongated with prominent sutures. Both the fruit apex and base are unusually blunt (Fig. 4).

Nut maturity in 'Schley' precedes that in 'Stuart' by about two days on the average. However, following mild winters, nut maturity can be about 10 to 14 days before 'Stuart.' In such cases, nut maturity in 'Schley' is not necessarily earlier than normal. Instead, nut maturity is delayed in 'Stuart' following a mild winter, because 'Stuart' has a greater chilling requirement than 'Schley' for bud break. The resulting delayed bud break in 'Stuart' subsequently delays nut maturity (21). Shuck dehiscence is much more uniform than in 'Stuart.' In some years, the shuck of some fruits open prematurely (Fig. 5). However, the disorder is minor compared to the severity that can occur in 'Success.' With premature shuck opening or "tulip disease," the kernel is under developed and the shell is soft or semi-pliable. In another minor disorder, all fruits in some clusters are undersized. This disorder also occurs in low frequency, and the fruits die on the tree before nut maturity.

Nut size and shape are variable on the same tree, and size varies from oblong to obovate. The base is mucronate, and the apex varies from mostly cuspidate to occasionally asymmetrical. The apex is grooved along the suture. Within nuts, shell halves are predominantly equal in size, but occasionally are unequal. In cross-section, the nut is oblong with the sides decidedly flattened. The suture is not elevated, and ridges are not evident. Shell topography is rough on the flattened sides but is smooth otherwise. The shell is moderately marked with long narrow stripes and fine dots. Markings are dark brown on a brown background (21).

Size averages about 65 nuts per pound. The shell is very thin, which can create two problems. First, the shell and shuck sometimes split. This is

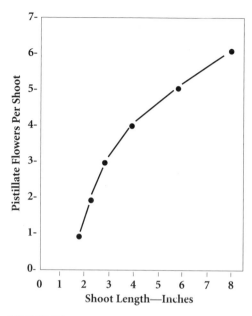

FIGURE 3.

Relationship of pistillate flowers per shoot to shoot length in 'Schley' pecan. Adapted from Gossard (12).

FIGURE 4.

'Schley' fruit photographed August 7, Philema, Georgia. Sutures are prominent; both base and apex are characteristically blunt.

FIGURE 5.
Premature shuck opening in 'Schley' (points). This disorder is sometimes referred to as "tulip disease." The disorder can occur in many other cultivars, and especially in years of high fruit set. Photographed October 12, 1988, Perry, Georgia.

the same disorder that occurs in 'Wichita.' Fruit split occurs at the maximum water stage of development, provided excessive soil moisture (as from heavy rain) and high humidity also occur at this stage of fruit development. The split occurs because of excessive pressure from the influx of water into the fruit. In southwest Georgia, fruit split in 'Schley' is neither as frequent nor as severe as in 'Wichita.' The reduced tendency of shuck split to occur in 'Schley' in comparison to 'Wichita' is due to the interrelationship of climatic conditions and fruit development. On the average, there is less rainfall during the water stage of 'Schley' and more rainfall during the water stage of the earlier maturing 'Wichita.' Second, if the nuts are harvested early (when the nut contains 20-30% moisture) and are shaken on a sunny day, the shell often splits while lying on the ground. The shell splits because it dries and shrinks faster than the kernel (21).

The 'Schley' nut is well adapted to mechanical shelling, so that a very high percentage of intact halves is obtained (30). This adaptability to mechanical shellings results from the thin shell, the thin and brittle central partition wall, and the shallow kernel grooves that readily release the packing tissue.

A percent kernel of 58 is good, with 60 being excellent. When scab is controlled and soil moisture is adequate, percentage kernel is very consistent from year to year. Kernel flavor is outstanding. Kernel color is very good (21) with a color rating of 8.4 (1 = dark; 10 = light).

The thin shell, the attractive and uniform kernel color, and the excellent flavor make 'Schley' nuts highly preferred by the consumer, but the slender kernel is objectionable to the confectioner. In addition, kernel color and color uniformity do not hold up well during storage. A black streak tends to develop on the dorsal surface of the kernel. Kernel stability is better when the nut is shelled than when unshelled (30), which contrasts with most cultivars. Furthermore, the stored kernel is reported to be very susceptible to pink rot (2). Consequently, 'Schley' nuts need to be marketed very soon after harvest, as is the common practice.

In the southeastern United States, scab is an economic factor in the production of 'Schley' Because of 'Schley's' high susceptibility to scab (4) and the difficulty of fungicide sprays penetrating the dense foliage, more fungicide sprays are needed than in less susceptible cultivars such as 'Stuart.' Consequently, production cost is increased. In Louisiana, 'Schley' is highly susceptible to bunch disease (15). The foliage is highly susceptible to fungal leaf scorch (16).

The fruits of 'Schley' have been noted for a long time (29) for their high susceptibility to injury by stink bugs and leaffooted bugs (10). In the southeastern United States, severe damage frequently occurs from these insects. Damage during the water stage of fruit development results in fruit abortion. This abortion is often confused with abortion from fruit split, which also occurs at this time.

'Schley' foliage is especially susceptible to damage by black pecan aphids (21, 25). In an orchard with mixed cultivars, early season infestations often can be controlled by spraying only the 'Schley' trees (W.L. Tedders, personal communication). The leaves are also especially susceptible to leafminers (9). 'Schley' is highly resistant to sooty mold accumulation following a black-margined aphid infestation (21, 23). The fruit is highly susceptible to damage by hickory shuckworm (28). Susceptibility to both southern pecan leaf phylloxera (6, 8) and pecan phylloxera (4) is moderate to high. Compared to 'Stuart,' 'Schley' is susceptible to spittlebug. 'Schley' fruits may be especially susceptible to abortion during a severe drought (11). In contrast, 'Schley' nuts tend to fill better under drought conditions than nuts of most other cultivars. The relatively small nut and thin shell make 'Schley' a favorite of bluejays, crows, and squirrels. 'Schley' is more resistant to zinc deficiency than 'Stuart.'

High quality 'Schley' nuts command a premium price, and, under good cultural practices, mature 'Schley' trees are profitable if nuts are harvested early and marketed rapidly. The more or less annual production of high quality nuts is the major advantage of 'Schley.' The major disadvantage is 'Schley's' high susceptibility to scab. Economics of controlling this disease will be the dominant factor governing the future success of this cultivar.

Literature cited

1) Alben, A.O., B.G. Sitton, E.N. Dodge, and O.W. Harris. 1953. Comparative growth and yield of 'Stuart,' 'Success,' and 'Schley' pecan varieties. *Proc. Amer. Soc. Hort. Sci.* 61:307-310.

2) Anon. 1952. Panel discussion on pecan varieties. *Proc. Southeastern Pecan Growers Assn.* 45:28-46.

3) Anon. 1962. Harris Super. Simpson Nursery Company, Monticello, Florida.

4) Anon. 1970. Dooryard pecan trees in Louisiana. Circular Letter to Homeowner. USDA Pecan Lab., Shreveport, Louisiana.

5) Anon. 1984. The Forkert pecan…Old but still popular. *Pecan South* 11(1):18-19.

6) Boethel, D.J., R.D. Eikenbary, and R.W. McNew. 1977. Phylloxera affects varieties differently. *Pecan Quarterly* 11(2):12-13.

7) Brooks, R.M. and H.P. Olmo. 1956. Register of new fruit and nut varieties: List 11. *Proc. Amer. Soc. Hort. Sci.* 68:611-631.

8) Calcote, V.R. 1983. Southern pecan leaf phylloxera *(Homoptera:Phylloxeridae)*: Clonal resistance and technique for evaluation. *Environ. Entomol.* 12:916-918.

9) Dutcher, J.D. and J.A. Payne. 1979. Bionomics and control of pecan leafminers. *Proc. Southeastern Pecan Growers Assn.* 72:143-146.

10) Dutcher, J.D. and J.W. Todd. 1983. Hemipteran kernel damage of pecan, pp. 1-11. *In:* J.A. Payne (ed.), Pecan Pest Management—Are We There? Misc. Pubs. *Ent. Soc. Amer.* 13(2).

11) Gammon, Jr., N., R.H. Sharpe, and R.G. Leighty. 1956. Relationship between depth to heavy textured subsoil and drought injury to pecan. *Proc. Southeastern Pecan Growers Assn.* 49:37-43.

12) Gossard, A.C. 1933. The importance of maintaining vigorous terminal growth on pecan trees. Proc. Natl. Pecan Assn. Bul. 32:84-89.

13) Hinrichs, H.A. 1962. Pecan industry and some research developments in Oklahoma. Ann. Rpt. Northern Nut Growers Assn. 53:80-85.

14) KenKnight, G.E. 1970. Pecan varieties "happen" in Jackson County, Mississippi. *Pecan Quarterly* 4(3):6-7.

15) KenKnight, G. and J.H. Crow. 1967. Observations on susceptibility of pecan varieties to certain diseases at the Crow farm in DeSoto Parish, Louisiana. *Proc. Southeastern Pecan Growers Assn.* 60:48-57.

High quality 'Schley' nuts command a premium price.

16) Littrell, R.H. and R.E. Worley. 1975. Relative susceptibility of pecan cultivars to fungal scorch and relationship to mineral composition of foliage. *Phytopathology* 65(6):717-718.

17) Madden, G.D. and E.J. Brown. 1975. Budbreak, blossom dates, nut maturity, and length of growing season of the major varieties grown in the West. *Pecan South* 2(3):96, 97, 112, 113.

18) Nakayama, R.M. 1967. Pecan variety characteristics. New Mexico Agr. Exp. Sta. Bull. 520.

19) O'Barr, R.D., W. Sherman, W.A. Young, W.A. Meadows, V. Calcote, and G. KenKnight. 1989. 'Moreland'—A pecan for Louisiana and the Southeast. Louisiana Agr. Exp. Sta. Circ. 129.

20) Reed, C.A. 1926. Historical pecan trees. Proc. Natl. Pecan Assn. Bull. 25:139-156.

21) Sparks, D. 1992. Pecan cultivars—The orchard's foundation. *Pecan Production Innovations.* Watkinsville, Georgia.

22) Sparks, D. and J.A. Payne. 1986. Notes on severity of tornado damage to 'Stuart' vs. 'Schley' pecan trees. Ann. Rpt. Northern Nut Growers Assn. 77:79-80.

23) Sparks, D. and I. Yates. 1990. Pecan cultivar susceptibility to sooty mold related to leaf surface morphology. *J. Amer. Soc. Hort. Sci.* 116(1):6-9.

24) Taylor, W.A. 1906. Promising new fruits. Pecans, pp. 504-508. *In:* G. W. Hill (ed.). *Yearbook of the USDA 1905,* Gov. Print. Off., Washington, D.C.

25) Tedders, W.L. 1978. Important biological and morphological characteristics of the foliar-feeding aphids of pecan. USDA Tech. Bul. 1579.

26) Thompson, T.E. and L.D. Romberg. 1985. Inheritance of heterodichogamy in pecan. *J. Hered.* 76:456-458.

27) USDA, Agr. Res Ser. 1955. Effects of the March 1955 freeze on peach, pecan, and tung trees in the South. ARS 22-18.

28) Walker, F.W. 1933. The pecan shuckworm. Univ. Florida Agr. Exp. Sta. Bull. 258.

29) Woodard, J.S., L.D. Romberg, and E.J. Willman. 1930. Pecan growing in Texas. Texas Dept. Agr. Bull. 95:151.

30) Woodroof, J.G. and E.K. Heaton. 1961. Pecans for processing. Georgia Agr. Exp. Sta. Bull. 80.

31) Worley, R.E., O.J. Woodard, and B. Mullinix. 1983. Pecan cultivar performance at the Coastal Plain Experiment Station. Univ. Georgia Agr. Expt. Sta. Res. Bull. 295.

32) Yates, I.E. and C.C. Reilly. 1990. Somatic embryogenesis and plant development in eight cultivars of pecan. *HortScience* 25:573-576.

Author Sparks was with the Department of Horticulture, University of Georgia, Athens, Georgia, when this article originally appeared, October 1994.

The 'Glen Moy' Red Raspberry

by Derek Jennings

Since its release from the Scottish Crop Research Institute in 1982, 'Glen Moy' has rapidly become the leading cultivar for early fruiting in Europe, and the cultivar most frequently chosen for all new plantings in Britain. British raspberry growers, particularly those in Scotland, have always preferred cultivars that crop early, partly because the raspberry season is later than in many other production areas. Hence, 30 years ago the most widely grown cultivars in Britain were 'Malling Promise' and 'Malling Jewel.' This was followed by 'Glen Clova,' and it is now 'Glen Moy;' these were all the earliest cultivars available. On average, 'Glen Moy' was two days earlier than 'Glen Clova' in a Scottish trial (1).

Recently, growers in Holland and France have realized the considerable market requirement for very early raspberries, and have devised means for producing early fruit under protection some two months before field production starts. 'Glen Moy' suits these methods well. Indeed, it could be said to be a cultivar which became available in the right place at the right time.

One reason why 'Glen Moy' has replaced 'Glen Clova' so rapidly is the ban on the use of dineseb-in-oil to control excessive vegetative vigor. The canes of 'Glen Clova' are difficult to manage unless they are chemically burned down in spring, but this is not the case for 'Glen Moy.' However, the more important reason for preferring 'Glen Moy' is its much superior fruit quality. It has an excellent flavor, pleasantly sharp with

> *It is fungal diseases, particularly root rot, that have provided the biggest threat to 'Glen Moy.'*

aromatic overtones; its fruits are uniformly large, typically averaging 3.5 to 4.0 grams and appreciably larger than those of 'Glen Clova.' This gives the cultivar a potential for much higher yields. It is probably the main reason why 'Glen Moy' has usually been a top yielder in trials. The fruits are also firm, have a good shelf-life and an excellent medium red color, though without the gloss preferred in parts of North America.

'Glen Moy' and 'Glen Prosen' were released at the same time and were the first red raspberries to be totally spine-free, being homozygous for gene *s*, which they have inherited from 'Burnetholm,' a distant ancestor. They are thus both grower-friendly and picker-friendly, and are particularly popular among pick-your-own growers and amateurs.

The pedigree of 'Glen Moy' is complex. It is a fourth backcross hybrid of the black raspberry, *R. occidentalis*, which probably contributed to the firmness of its fruit texture. One of its grandparents is 'Glen Clova,' which probably contributed to its earliness, though several of its other ancestors were also early.

Methods for producing very early strawberries under protection have been perfected for sometime in continental Europe, and it is only in recent years that attempts are being made to achieve the same success with raspberries. The methods require an early cultivar to give maximum response to the early rise in spring temperature, and 'Glen Moy' has proved ideal. A small proportion of 'Glen Clova' is sometimes grown with it because of this cultivar's reputation for copious pollen production. A common method is to plant raspberry canes in mobile containers (buckets, baskets or plastic grow-bags) in October or November, hold them in a cold store until January or February, and then move them to plastic tunnels or glasshouses in January or February for cropping in late April and May. The containers are then moved to provide space for another crop. Sometimes the canes are densely planted in a plastic tunnel and removed after cropping. Yields of 1.0 to 1.5 kilograms per square meter are claimed with an average fruit weight of 8 grams.

'Glen Moy' rarely suffers from virus disease. It has major-gene resistance to two of the four strains of the large European aphid and virus vector *(Amphorophora rubi)*, but it commonly carries small populations of this aphid. Like 'Glen Clova,' it carries gene *Ls*, which gives it a highly sensitive, frequently lethal reaction when infected with raspberry leaf spot virus (4). Most cultivars are tolerant of this virus and carry it without symptoms; they therefore provide a source of inoculum to infect sensitive ones. In 'Glen Moy,' the disease tends to be limited to a few plants on the periphery of plantations and does not spread as extensively as in a more aphid-susceptible cultivar such as 'Glen Clova.' More importantly, the cultivar is immune from the common strains of the pollen-borne raspberry bushy dwarf virus and consequently shows neither crumbly fruit nor the reduction in vigor commonly associated with infection by this virus (5).

It is fungal diseases, particularly root rot caused by *Phytophthora fragariae* var *rubi*, that have provided the biggest threat to 'Glen Moy.'

Few cultivars have strong resistance to this pathogen, but 'Glen Moy' is particularly susceptible. It is unfortunate that the cultivar was launched in the 1980s when the disease first became serious and spread with such devastation in Britain and Europe. Prospects for the future look better, because propagators now take more precautions to ensure that the disease is not unwittingly spread in the planting material, and more effective chemicals are available to control it in fruiting plantations.

'Glen Moy' is also particularly susceptible to midge blight, a disease complex caused by the raspberry midge, *Resseliella theobaldi*, in combination with a range of fungi. The midge lays its eggs in natural splits that occur in the rind of raspberry canes, and 'Glen Moy' is vulnerable because its canes produce an abundance of splits early in the season. The feeding sites of the emerging larvae provide gaps in the periderm through which several pathogens which cannot invade intact stems are able to infect. These cause extensive death of the fruiting canes, but, unlike diseases caused by root-rot fungi, the new growth is vigorous and healthy, and growers have the means to control the midge and restore affected plantations to full production.

The third fungal disease to which 'Glen Moy' is highly susceptible is yellow rust *(Phragmidium rubi-idaei)*. This disease occurs too late in the season to cause serious damage in Britain, but it would probably cause premature defoliation if it occurred earlier, as it does in some parts of the world.

On the credit side, 'Glen Moy' has good resistance to spur blight *(Didymella applanata)* and cane botrytis *(Botrytis cinerea)* (3), and has a degree of resistance to *Leptosphaeria coniothyrium* (2), which causes cane blight. These resistances are thought to be associated in part with the pubescence and spinelessness of the canes. The fruits show a degree of resistance to grey mold *(B. cinerea)* and good resistance to powdery mildew *(Sphaerotheca macularis)*, which, together with their good texture, gives them a good shelf-life.

'Glen Moy' is owned by Plant Breeding International of Cambridge, United Kingdom, and is protected by Plant Breeders Rights.

Literature cited

1. Cormack, M.R. and S.L. Gordon. 1990. Raspberry cultivar trial 1980-1985. Occasional Pub. No. 9, Scot. Soc. for Crop Res.

2. Jennings, D.L. and E. Brydon. 1989. Further studies on resistance to *Leptosphaeria coniothyrium* in the red raspberry and related species. *Ann. Appl. Biol.*, 115:499-506.

3. Jennings, D.L. and E. Brydon. 1989. Further studies on breeding for resistance to *Botrytis cinerea* in red raspberry canes. *Ann. Appl. Biol.*, 115:507-513.

4. Jones, A.T. and D.L. Jennings. 1980. Genetic control of the reactions of raspberry to black raspberry necrosis, raspberry leaf mottle and raspberry leaf spot viruses. *Ann. Appl. Biol.*, 96:59-65.

5. Jones, A.T., A.E Murant, D.L. Jennings and G.A. Wood. 1982. Association of raspberry bushy dwarf virus with raspberry yellows disease; reaction of *Rubus* species and cultivars, and the inheritance of resistance. *Ann. Appl. Biol.*, 100:135-147.

Author Jennings was with the Scottish Crop Research Institute (now Clifton), Otham, Maidstone, United Kingdom, when this article originally appeared, January 1995.

The 'Western Schley' Pecan

by Darrell Sparks

'Western Schley' is by far the most widely planted cultivar in the western United States pecan belt. 'Western Schley' was originated by E.E. Risien, San Saba, Texas. This cultivar was selected by Risien from his orchard of some 1,000 San Saba seedlings (6). Risien planted the seed for this orchard in 1895 (26). The nuts were planted with the belief that many of the seedlings would come true to their parent (10), which, of course, did not happen. The selection was named 'Western Schley' by Risien and released by Risien sometime before or in 1924 (2). Evidently, Risien believed this selection was equal to 'Schley, which was grown primarily in the southeastern United States, hence the name, 'Western Schley.' 'Western Schley' has been used in breeding, but only one cultivar, 'Harper,' has been released with 'Western Schley' parentage. Bud break is relatively late and occurs about the same time as 'Stuart' (24). The foliage is glossy and exceptionally dark green. The color fades under conditions of prolonged exposure to intense sunlight. Fading, however, does not occur as readily as in 'Wichita' and 'Schley.' The leaflets are especially wavy or convoluted as in 'Wichita' and Mahan.' The distal portion of the rachis often curves upward, with the result that the leaf appears to be swayback as in 'Curtis.' The shoots tend to be spindly with a relatively high number of leaves per unit of length. In the early spring, the bark of the shoot is bright green and glossy. As the growing season progresses, the bark turns reddish brown. Sometimes, older branches have patches of bark with reddish iron-like color. The coloration of both the shoot bark and branch bark are distinguishing characteristics of 'Western Schley.' The tree has a spreading canopy with an open growth form, which is enhanced by the pole-like branches (Fig. 1). This open canopy no doubt contributes to a high production of nuts. The tree is a very vigorous grower. Like 'Stuart,' the crotch angles of 'Western Schley' are relatively narrow but strong. Hence, the tree requires very little training.

'Western Schley' is protandrous. 'Ideal' (mistakenly called 'Bradley') is a reasonably good pollinizer for 'Western Schley' in the Las Cruces pecan region of New Mexico (24). Extensive overlapping of pollen shedding and pistillate receptivity often occurs in 'Western Schley' (7, 11, 13, 16, 19, 28). For this reason, selfing is commonly assumed to occur in this cultivar. Evidence for the capacity for selfing in this cultivar is the substantial fruit set occurring in an isolated 'Western Schley' orchard planted without pollinizers near Lubbock, Texas. Additional indication of selfing is the abnormally heavy third fruit drop in this orchard. Selfing has been demonstrated to increase the magnitude of the third drop (20, 25). Selfing also reduces kernel quality (20), and, consequently, extensive selfing in 'Western Schley' would be expected to decrease kernel development. That the kernel is detrimentally affected by selfing is supported by the fact that cross-pollination increases kernel quality in 'Western Schley' (14). 'Western Schley' seed make poor rootstocks in that the seedlings are small, as demonstrated by Hinrichs (12) and by data of Mielke, et al. (15). The small size of seedlings has been attributed to extensive selfing in 'Western Schley.' In the Crystal City, Texas, area, 'Western Schley' often has a relatively large fourth drop.

'Western Schley' is moderately precocious (24), and, under excellent cultural conditions, the tree can be in commercial production within six years from transplanting. The fruit is borne on shoots from relatively short branches that arise along the length of long branches. Thus, as in 'Pawnee' and 'Shawnee,' the fruit appears to be borne along the length of long poles. Fruiting shoots tend to be maintained along the branches because of the open canopy. When in heavy production, these poles often droop from fruit weight in July-September (Fig. 2). 'Western Schley's' fruiting pattern makes the tree particularly suited to hedging. When the tree is heavily hedged, nut production is restored within three to four years. The tree is prolific and, as a result, very well-managed 'Western Schley' orchards with a full stand of trees may have average yields of 1,500 to 2,000 pounds per acre. The tree is a fairly consistent producer.

Date of nut maturity is about three days before 'Stuart' (24). Under some conditions, the shuck dries along the suture prior to dehiscence (Fig. 4). The shuck is thick (5) with pronounced sutures (Fig. 3).

Nut shape is obovate. Base shape is mostly obtuse, but is variable and sometimes is obtuse asymmetric, round asymmetric, or acute. The apex is likewise very variable. Although mostly cuspidate asymmetric, the apex can also be cuspidate or obtuse asymmetric. The apex is broadly grooved along the suture. Most nuts have unequal shell halves. Nut cross-section is oblate.

FIGURE 1.
Tree form is 'Western Schley.' Note the pole branching habit which makes this cultivar more suited to mechanical hedging than most other pecan cultivars.

FIGURE 2.
Drooping branch habit of 'Western Schley' pecan trees due to the formation of fruit on short branches along the length of a long slender pole-like branch.

FIGURE 3.
Spindly peduncle of 'Western Schley' fruit cluster makes mechanical shaking of the nuts difficult.

FIGURE 4.
Drying along the suture of 'Western Schley' fruit just prior to shuck opening. This characteristic is not universal on all fruits.

The suture is elevated, but ridges are not evident. Shell surface is exceptionally rough. Stripes are moderate. Much of the shell surface is densely dotted, causing the nut to appear dirty and unattractive. Markings are reddish-brown on a light brown background (24).

Nuts are not large and average about 64 nuts per pound. The nut is too small to sell well in the in-shell trade. The shell is thin. A percentage kernel of 57 is good with 60 being excellent. The kernel color rating is 6.2 (1 = dark; 10 = light), which is somewhat less than 'Stuart' and 'Desirable.' Mature 'Western Schley' trees tend to overproduce, with the result that the kernel percentage decreases down to the low 50s. Thus, the capacity of 'Western Schley' to produce high quality nuts is greatly diminished.

The tendency for kernel quality to decrease with tree age is a major disadvantage. Kernel quality can be restored by hedging because hedging increases the leaf area per fruit. Hedging to maintain nut quality has been practiced extensively in the El Paso, Texas-Las Cruces, New Mexico area; but the practice is on the decline, mainly because of the loss in nut production during the first three years or so following hedging (24).

Although a high percentage of the 'Western Schley' crop is shelled, the nut is not a good cracker. The difficulty of removing packing tissue from the deep kernel grooves during shelling and the likelihood for the kernel shoulders to break during cracking decrease the adaptability of 'Western Schley' to mechanical processing. The shoulders tend to break because the rigid, central partition wall does not freely separate from the shell during cracking and shelling (24). Kernel stability either in-shell or shelled is only fair (27).

'Western Schley' has one characteristic that is especially favorable for climates that are marginally short for pecan; for example, the El Paso area of Texas, the Las Cruces area of New Mexico, and Oklahoma. This characteristic is that the tree enters dormancy with ensuring leaf drop relatively early in the fall. Early seasonal dormancy is of primary importance in marginal climatic areas in order for the tree to escape major damage from freezes. Young 'Western Schley' trees also have good resistance to fall freezes in the southeastern United States (9).

'Western Schley' is very susceptible to scab (17, 18, 21). However, it can be grown successfully under the humid conditions of the southeastern United States if properly timed fungicide sprays are applied. The open canopy is no doubt a contributing factor in scab control. Regardless, 'Western Schley' is not recommended for the southeastern United States because the small nuts are not suited for the Thanksgiving and Christmas trade. 'Western Schley' is extremely susceptible to downy spot (8), and, with severe infections, massive defoliation occurs.

The foliage is also susceptible to vein spot and liver spot (8) and is moderately susceptible to fungal leaf scorch. 'Western Schley' is immune to southern pecan leaf phylloxera (1, 3), but is susceptible to pecan phylloxera (4). 'Western Schley' has moderate resistance to hickory shuckworm (5). 'Western Schley' foliage is very susceptible to potato leafhopper damage. The damage can be as severe or worse than with 'Desirable,' which is also very sensitive to this insect. The susceptibility of 'Western Schley' to yellow and black-margined aphids is acute, with nearly the severity of 'Cheyenne.'

One of the most critical nutritional parameters for 'Western Schley' tree growth and fruit production is a proper balance of nitrogen and potassium. Developing fruits differentially drain potassium from the leaves (22), which may cause a high nitrogen to potassium ratio or accentuate an existing high ratio. If fruiting is excessive and leaf potassium is very low, the tree dies back during the following growing season and, in extreme cases, may die. Dieback and tree death have been major problems with 'Western Schley' in Brazil. The impact of an imbalance of nitrogen and potassium is more severe for 'Western Schley' than 'Desirable' due to 'Western Schley's' higher fruiting capacity. This cultivar is tolerant to low zinc (10). 'Western Schley' is much more salt-tolerant than 'Wichita' and, consequently, does better than 'Wichita' under saline conditions. The third fruit drop is especially sensitive to drought (23).

The nuts on a 'Western Schley' tree are difficult to shake mechanically, particularly if harvested early. This difficulty is due to the fact that the fruit is borne on a whipped shoot with a long spindly peduncle (Fig. 4). When the tree is shaken, the net result is that the shaking energy is dissipated in the whippy shoot rather than in the fruit cluster. 'Western Schley' nuts are especially susceptible to losses from birds. The slender nut is easily carried by birds.

Overall, 'Western Schley' is not an ideal cultivar. However, 'Western Schley's' high yielding ability and suitability to marginal pecan climates make this cultivar highly successful. The huge acreage of relatively young trees will ensure 'Western Schley's' influential position in the market for many years; but, as these trees mature, maintaining nut quality will become an increasing problem.

Literature cited

1) Boethel, D.J., R.D. Eikenbary, and R.W. McNew. 1977. Phylloxera affects varieties differently. *Pecan Quarterly* 11(2):12-13.

2) Burkett, J.H. 1924. The pecan in Texas. Texas Dept. Agr. Bull. 77.

3) Calcote, V.R. 1983. Southern pecan leaf phylloxera *(Homoptera: Phylloxeridae):* Clonal resistance and technique for evaluation. Environ. Entomol. 12:916-918.

4) Calcote, V.R. and D.E. Hyder. 1980. Pecan cultivars tested for resistance to pecan phylloxera. *J. Georgia Entomol. Soc.* 15: 428-431.

5) Calcote, V.R., G.D. Madden, and H.D. Peterson. 1977. Pecan cultivars tested for resistance to hickory shuckworm. *Pecan Quarterly* 11:4-5.

6) Crane, H.L., C.A. Reed, and M.N. Wood. 1938. Nut breeding, pp. 827-887. *In:* G. Hambige (ed.). *Yearbook of Agriculture 1937.* Gov. Print. Off., Washington, D.C.

7) Da Costa Baracuhy, J.B. 1980. Determinagao do periodo de floragao e viabilidade do pollen de differentes cultivares de noqueira pega, *Carya illinoensis* (Wang.) K. Koch. Ph.D. Diss. Universidade Federal De Pelotas Pelotas, Brazil.

8) Dodge, F.N. 1944. Pecan varieties. *Proc. Texas Pecan Growers Assn.* 23:7-16.

9) Goff, W.D. and T.W. Tyson. 1991. Fall freeze damage to 30

genotypes of young pecan trees. *Fruit Var. J.* 45:176-179.

10) Gray, O.S. 1974. Looking at the outside and inside of some pecan varieties. *Proc. Western Pecan Conf.* 8:73-81.

11) Hinrichs, H.A. 1962. Pecan industry and some research developments in Oklahoma. *Annu. Rpt. Northern Nut Growers Assn.* 53:80-85.

12) Hinrichs, H.A. 1965. Pecan investigations in Oklahoma. *Annu. Rpt. Northern Nut Growers Assn.* 56:44-51.

13) Madden, G.D. and E.J. Brown. 1975. Here are methods to improve pollination. *Pecan Quarterly* 9(4):10-12.

14) Marquard, R.D. 1988. Outcrossing rates in pecan and the potential for increased yields. *J. Amer. Soc. Hort. Sci.* 113:84-88.

15) Mielke, E.A., H. Tate, and D. Bach. 1978. Seedling pecan rootstock for size controls: A preliminary report. *Proc. Western Pecan Conf.* 12:7-17.

16) Nakayama, R.M. 1967. Pecan variety characteristics. New Mexico Agr. Exp. Sta. Bull. 520.

17) Romberg, L.D. 1931. Most desirable varieties for Texas, and reasons. *Proc. Natl. Pecan Assn.* 30:35-43.

18) Sanderlin, R.S. 1987. Evaluation of pecan cultivars and USDA selections for scab disease susceptibility. Res. Rpt. Pecan Res. Exp. Sta., Louisiana Agr. Exp. Sta., Shreveport, Louisiana.

19) Sibbett, G.S., T.E. Thompson, and N. Troiani. 1988. Pecans for California. *Pecan South* 22(3):16-21.

20) Smith, C.L. and L.D. Romberg. 1940. Stigma receptivity and pollen shedding in some pecan varieties. *J. Agr. Res.* 60:551-564.

21) Sparks, D. 1976. Scab control notes for 1974 and 1975. *Pecan South* 3(1):290-291.

22) Sparks, D. 1977. Effects of fruiting on scorch, premature defoliation, and nutrient status of 'Chickasaw' pecan leaves. *J. Amer. Soc. Hort. Sci.* 102:669-673.

23) Sparks, D. 1988. Drought stress induces fruit abortion in pecan. *HortScience* 24:78-79.

24) Sparks, D. 1992. Pecan cultivars—The orchard's foundation. *Pecan Production Innovations*. Watkinsville, Georgia.

25) Sparks, D. and G.D. Madden.1985. Pistillate flower and fruit abortion in pecan as a function of cultivar, time, and pollination. *J. Amer. Soc. Hort. Sci.* 110:219-223.

26) Taylor, W.A. 1908. Promising new fruits. Pecans, pp. 315-320. *In:* J.A. Arnold (ed.). *Yearbook of the USDA, 1907.* Gov. Print. Off., Washington, D.C.

27) Woodroof, J.G. and E.K. Heaton. 1961. Pecans for processing. Georgia Agr. Exp. Sta..

28) Worley, R.E., O.J. Woodard, and B. Mullinix. 1983. Pecan cultivar performance at the Coastal Plain Experiment Station. Univ. Georgia Agr. Exp. Sta. Res. Bull. 295.

Author Sparks was with the Department of Horticulture, University of Georgia, Athens, Georgia, when this article originally appeared, April 1995.

The 'Senga Sengana' Strawberry

by Edward Zurawicz & Hugh Daubeny

'Senga Sengana' has had remarkable longevity for a strawberry cultivar. It was released in 1954 and still is the major cultivar grown throughout central and eastern Europe and in the Scandinavian countries (1, 7). The continuing importance of 'Senga Sengana' can be appreciated by the fact that it still accounts for 80% of the strawberries produced in Poland, the country second to the United States in world production. Between 70 and 80% of 'Senga Sengana' fruit is processed.

'Senga Sengana' has a fascinating history. It was released from the program of Dr. Rudlof von Sengbusch, who began breeding strawberries in 1942, during the second world war, at the State Experiment Station at Luckenwalde, near Berlin, Germany (3). Dr. von Sengbusch had established his reputation as a breeder of rye, hemp, spinach, asparagus, and lupin. The first alkaloid-free lupin cultivar was among his accomplishments. The decision to breed strawberries was based on the needs of the newly developed deep-freeze industry to have cultivars producing fruit suited to its purposes. None of the cultivars grown at that time in Germany met all of the industry requirements (9). The program was supported by the processing companies.

Dr. von Sengbusch screened all available cultivars for adaptation. Only one, 'Markee,' which was of American origin but named after the village where it was first grown in Germany, had any sort of adaptation (3, 9). However, the cultivar was quite unproductive and the fruit without aroma. It was subsequently used in crosses with various other cultivars, mostly of

European origin, which were noted for greater productivity and produced more aromatic fruit. Approximately 40,000 seedlings from these crosses were planted in the field in 1944. In the following years, more than 75% of these were discarded because of poor vigor; fruits of those remaining were tested for deep-freezing suitability. From these, approximately 1,500 were propagated and established in test plots at three locations, each of which, like Luckenwalde, was in East Germany, at that time occupied by the Russian army. In 1948, the entire breeding program was relocated to Wulsdorf, near Hamburg, Germany.

In 1949, the selection fields at Wulsdorf were severely infested with cyclamen (strawberry) mite, *Phytonemus pallidus* (Banks) (*Tarsonemus pallidus* Zimm.). All but three selections were badly stunted by the infestations of the pest. The three selections originated from the cross of 'Markee' x 'Sieger.' 'Sieger' was an old German cultivar, released in 1987 from the cross of 'Kaiser Sampling' x 'Laxton's Noble.' One of the three was particularly productive, with fruit considered ideal for freezing and other forms of processing. In 1954, the selection was introduced for commercial production in Germany under the name 'Senga Sengana' (3, 9). At the same time, von Sengbusch founded the Sengana GmbH. company and assumed the position of president.

According to the original description, 'Senga Sengana' was midseason ripening with a comparatively short harvesting period lasting from 15 to 20 days (9). The relatively firm fruit, which maintained its integrity upon thawing, was described as having glossy, dark carmine red external and internal color, a heart to round shape, large to medium size, easily removed calyx, and a sweet, mild flavor. All these traits were considered essential for fruit suited to processing. In addition, the fruit seemed suited to the local German fresh market, especially if harvested before full ripe without removing the calyx. The plants were described as high yielding with stable production from year to year.

Soon after its release, 'Senga Sengana' became the predominant cultivar in most European countries, including Germany, Belgium, Holland, Denmark, Norway, Sweden, Finland, Switzerland, Czechoslovakia, and Poland, and also throughout much of the then Soviet Union, including Estonia, Latvia, Lithuania, White Russia, and the Ukraine. The cultivar was a major factor in the increase in strawberry production in all of these countries. For example, by 1978, Poland produced 200,000 tons of strawberries per year; most of this was from 'Senga Sengana.' This enabled the country to become one of the world's largest exporters of deep-frozen strawberry fruit, as well as products such as juice, juice concentrate, purée, compotes. and precooled fruit (10).

'Senga Sengana' proved to be productive in narrow matted rows, giving yields of 15 to 20 metric tonnes of fruit per hectare in the first year of production. Higher yields were often obtained in the second year. Plants remained vigorous, and many growers were able to maintain plantings for as many as four years. Yields of other cultivars, including many introduced from North American breeding programs, did not equal those obtained from 'Senga Sengana' (1, 10). Initially described as ripening midseason, it was subsequently described as mid to late season (10).

Although suited to the local fresh market, with an excellent fresh dessert quality, 'Senga Sengana' fruit proved to be unsuitable for long-distance transport because it lacked a sufficiently high level of firmness and was susceptible to rot, caused by *Botrytis cinerea* Pers. ex. fr. (1, 4, 5). In recent years, the fruit has been further discounted for fresh market because it is smaller than that of newer cultivars. In western Europe, in particular, 'Senga Sengana' has been replaced by cultivars, such as 'Elsanta,' which are better suited to long-distance transport.

'Senga Sengana' is relatively resistant to two-spotted spider mite, *Tetranchus urticae* Koch. Although initially selected as showing tolerance to cyclamen mite, it will support relatively high populations of the pest. It is also tolerant to Verticillium wilt (*Verticillium albo-artum* Reinke & Berth).

The cultivar is relatively susceptible to preharvest rot, caused by *B. cinerea* as well as to postharvest rot caused by the same organism (1, 4, 5). Its preharvest susceptibility is probably influenced by the fact that most of the fruit are below the leaf canopy. 'Senga Sengana' is also susceptible to leaf spot (*Mycospaerella fragariae* [Tul.]) and to powdery mildew (*Sphaerotheca macularis* [Fr.] Magn.). However, in Russia, it was reported resistant to the latter (4). This discrepancy might be due to the occurrence of different races of the causal organism. The cultivar is susceptible to red stele root rot, caused by *Phytophthora fragariae* Hickman (6). Other susceptibilities include reactions to two species of leaf and bud nematodes, *Aphelenchoides fragariae* (Ritz. Boz. [Christ.]) and *Aphelenchoides rizemabosi* ([Schwartze] Steiner).

'Senga Sengana' is well adapted to the soil and climatic conditions of eastern, central, and northern Europe. It has also been grown successfully in the Atlantic provinces of Canada (2). Plants grow well at relatively low soil temperatures. This has been attributed to their strong and deep root systems; roots reach the depth of two meters in established plantings. Plants usually do not show winter damage, even when snow covering is lacking. However, in some exceptionally cold winters, with little or no snow cover, some crown damage can occur (7). Since flowering occurs relatively late and the inflorescences are below the leaf canopy, there is seldom damage to flowers or to flower buds from spring frosts.

The cultivar is an abundant runner producer. Thus, it is ideally suited to the matted row culture system. At same time, it is easily propagated.

There were many other cultivars released from Dr. von Sengbusch's breeding program, including 'Senga 29,' 'Senga 54,' 'Senga 188,' 'Senga Dulcita,' 'Senga Fructana,' 'Senga Gourmella,' 'Senga Jurica,' 'Senga Pantagruella,' 'Senga Precosana,' 'Senga Litessa,' 'Senga Tigajga,' and 'Senga Gigana' (3). None had the long-lasting impact on the strawberry industries of Europe that 'Senga Sengana' has had. The cultivar has also had an impact in many breeding programs. It appears in the derivations of

Soon after its release, 'Senga Sengana' became the predominant cultivar in most European countries.

cultivars such as 'Kama,' 'Solveta,' 'Kristina,' 'Lina,' 'Melody,' 'Dagmar,' 'Fratina,' 'Frignetta,' 'Hiku,' and 'Bounty.' These have, in turn, been used as parents.

It is likely that 'Senga Sengana' will remain an important cultivar well into the twenty-first century. This will make its life span even more remarkable and one that no other modern-day cultivar can come close to matching.

Literature cited

1) Cieslinski, G., A. Klimeczak, and K. Smolarz. 1993. The performance of some new American and Polish strawberry cultivars grown in Poland. *Acta Horticulturae* 348:171-176.

2) Craig, D.L., J.A. Cutcliffe, and W.B. Collins. 1971. Strawberry cultivar testing in the Atlantic provinces. *Fruit Var. Hort. Dig.* 25:83-85.

3) Darrow, G.M. and W.H.J. Hondelmann. 1966. Strawberry breeding and industry on the European continent. pp 270-300. *In:* Darrow, (ed.) *The Strawberry. History, Breeding and Physiology.* Holt, Rinehart and Winston, New York.

4) Govorova, G. 1993. Strawberry breeding in Russia. *Acta Horticulturae.* 348:45-55.

5) Kolbe, W. 1973. Untersuchungen uber den Einfluss der *Botrytis*-bekampfung auf die Sortenleistung in Erdbeeranbau in Abhangigkeit von der Witterungsbedingungen. *Erwerbsobstbau.* 15:40-45.

6) Kronenberg, H.G., C.P.J. van de Lindeloof, and L.M. Wassenar. 1971. Strawberry breeding for red core resistance in the Netherlands. *Euphytica.* 20:228-234.

7) Rosati, P. 1991. The strawberry in Europe. pp. 27-33. *In:* A. Dale and J.J. Luby (eds.) *The Strawberry into the 21st Century.* Timber Press, Inc., Portland, Oregon.

8) Smolarz, K. and E. Zurawicz. 1979. Comparative studies on the bearing of American, European, and the hybrids of strawberries obtained by the authors in the climatic conditions of central Poland. Prace Inst. Sad., Seria A, Tom 21.

9) Von Sengbusch, R. 1954. Erdbeersorte 'Senga Sengana,' ein neuer robuster Massentrager. Obstbau 73:105.

10) Zurawicz, E. and K. Smolarz. 1980. Strawberry production trends in continental Europe. pp. 41-52. *In:* N.E Childers (ed) *The Strawberry Cultivars to Marketing.* Horticultural Publications, Gainesville, Florida.

Authors Zurawicz and Daubeny were with the Research Institute of Pomology and Floriculture, Skierniewice, Poland; and the Research Station, Agriculture Canada, Vancouver, respectively, when this article originally appeared, July 1995.

The 'Fuji' Apple

by Yoshio Yoshida, Xuetong Fan, & Max Patterson

In 1939, the first apple breeding program was established at the Morioka Fruit Tree Research Station as a program of the Japan Ministry of Agriculture, Forestry and Fisheries. In 1958, Tohoku No. 7 was selected from 596 fruit-bearing hybrids of 'Ralls Janet' (female) and 'Delicious' (male). This selection was named 'Fuji' and registered by the Ministry of Agriculture and Forestry of Japan in 1962. The name 'Fuji' was a commemoration of the town of Fujisaki in the Aomori Prefecture where the crosses and selections were made. The original tree (Fig. 1) is still standing in the orchard of the Morioka Branch, Fruit Tree Research Station. The original strain is not highly colored, but many red coloring strains have been found. 'Fuji' fruits are medium to large size, firm, crisp, very sweet, and very juicy. Fruits develop slightly roughened skin, have a subacid favor, and are very good dessert apples. Fruit is round-oblate or oblong, and mature in early or mid-November in Morioka and often have watercore (Fig. 2). Trees are large, spreading, vigorous, and productive, but susceptible to biennial bearing.

Cultivation

Fruit quality and color of 'Fuji' on M.9 were superior to 'Fuji' on *M. prunifolia* when trees were grown in a volcanic ash soil (20). 'Fuji' on M.9 fruited one or two years earlier than trees on *M. prunifolia*. The tendency toward biennial bearing was less with M.9 than with *M. prunifolia*.. Yield of 'Fuji' on M.9 was higher than that on M.26. Fruit

FIGURE 1.
The original seedling tree of 'Fuji' (Tohoku No.7) at the Morioka Branch, Fruit Tree Research Station in Japan.

size, soluble solids, and flower bud formation tend to decrease as trees mature (20). 'Fuji' trees can become biennial bearing if overcropped. Early fruit thinning to one fruit per five terminal buds and 75 leaves per fruit resulted in adequate return bloom and good fruit size. 'Fuji' is difficult to thin, even with young trees. Ethephon sprays at full bloom give satisfactory thinning for 'Fuji' (10), but cause excessive russeting.

Maturation

As harvest is delayed, fruit weight and soluble solids of 'Fuji' apples increase, and firmness, titratable acids, starch, and total pectin decrease (12). During a seven-day after-harvest period at 20°C (68°F), soluble solids increased and titratable acid decreased, but firmness was unchanged. The slow rate of firmness loss by 'Fuji' apples is in contrast to rapid rates of loss exhibited by many other apple varieties (23). Groundcolor and red surface color as determined by color charts have been used as maturity indices in Japan (22). 'Fuji' apples have low ethylene evolution, internal ethylene, and low respiration rate during ripening (11, 23). Although 'Fuji' apples contained appreciable amounts of 1-aminocyclopropane1-carboxylic acid (ACC) just after harvest, ethylene evolution was not detected (23).

Storage

Controlled atmosphere (CA) storage (0-1°C [32-34°F], 3-4.1% O_2, and 1.5-2.8% CO_2 for 187 days) of slightly unripe 'Fuji' was successful for long-term storage (17). Internal browning was observed after six months of cold storage with greater incidence in more mature apples (17). Fruit stored in regular cold storage have greater weight loss than fruit stored in CA. Controlled Atmosphere storage also maintains acidity in 'Fuji.' 'Fuji' remained in a firm ripe or crisp condition, even after a warm treatment at 20°C (68°F) for 30 days, while other cultivars became overripe and mealy (6). Fruit in 1% of CO_2 had less internal browning than fruit in 3% CO_2. Early harvested 'Fuji' had superficial scald after seven months cold storage at 0°C (32°F) (4), but fruit stored at 0°C for eight months in low pressure storage at 40 or 80 mmHg maintained good quality (16). Fruit stored at 725 and 160 mmHg showed core browning after long-term condition after two or three months at 10°C (50°F) (18).

Diseases and Disorders

Watercore: 'Fuji' develops severe watercore if harvested late. Since Asian consumers like watercored apples, these fruits command high prices. Watercored fruits do not store well due to internal breakdown. A single beam of 810 nm light can be used to determine the incidence of watercore (7, 8). Watercore appeared in both the core and the flesh area of 'Delicious' but only developed in the core area of 'Fuji' (8).

Core browning: 'Fuji' apples have peculiar browning disorders referred to as "core-line browning" and "innercore browning" (5). The incidence of core-line browning tends to increase with increasing watercore. Watercore of 'Fuji' remains longer in the core-line region than in other areas. The incidence of inner-core browning was not related to watercore intensity. However, a relation between the occurrence of this disorder and watercore was suspected, because the inner core area is the last part which is affected by watercore during storage. In a Japanese study, preharvest calcium sprays caused an increase in core-line browning and scald (5).

'Fuji' apples are susceptible to several other diseases and disorders. Russet ring is a graft-transmissible virus (21) reported in Japan. Temperature affects symptom expression, with daytime temperatures higher than 20°C (68°F) resulting in suppression of symptoms. Apple fruit crinkle, also graft-transmissible, showed symptoms on 'Fuji' after dry fall weather (13). 'Fuji' is very susceptible to bitter rot (*Glomerella cingulata* Stoneman) (3). Cork spot, bitter pit, and a vanadate-sensitive ATPase related to bitter pit were found in 'Fuji' (9). 'Fuji' leaves seemed to have less susceptibility to *Alternara mali* Robert than 'Starking Delicious' (24). 'Fuji' is also very susceptible to mucor rot caused by *Mucor piriformis* (14).

Commercialization

'Fuji' quickly became an important cultivar in Japan and its culture was associated with the dramatic decline in acreage of 'Ralls Janet' and 'Jonathan.' 'Fuji' has also become a major variety in Korea and Brazil (1). Although introduced to the United States in the early 1960s, 'Fuji' has achieved wide popularity only in recent years. In the last five years, about 619,000 'Fuji' trees have been planted, and a considerable number of existing trees have been top-worked to 'Fuji' in

The slow rate of firmness loss by 'Fuji' apples is in contrast to rapid rates of loss exhibited by many other apple varieties.

Washington State (2). It is estimated that another 2.21 million 'Fuji' trees were sold in 1991 and 1992, and that the production of 'Fuji' in Washington will approach several million boxes by the end of the century. California will have 8,000 to 10,000 acres of 'Fuji' by 1995. California production could reach seven million boxes in the next several years (15). These production increases will make 'Fuji' one of the major apple varieties in the United States.

Literature cited

1) Bernardi, J. 1988. Behavior of some apple cultivars in the subtropical region of Santa Catarina, Brazil. *Acta Hort.* 232:46-50.

2) Buckner, L.R. 1991. Production, planting trends vital to planning decisions. *Good Fruit Grower* 42(4):28-31.

3) Camilo, A.P., R.C. Lamb, and H.S. Aldwinckle. 1988. Genetic resistance to bitter rot incited by *Glomerella cingulata* (Stoneman). Spaulding & Von Schrenk in apple [*Malus domestica* (Borkh.)]. *Acta Hort.* 232:37-42.

4) Cantillano, R.F.F. 1988. Time of harvesting and diphenylamine treatment in the control of superficial scald in apple cv. Fuji, in Rio Grande Do Sul, Brazil. *Acta Hort.* 232:245-252.

5) Fukuda, H. 1984. Relationship of watercore and calcium to the incidence of internal storage disorders of 'Fuji' apple fruit. *J. Japan. Soc. Hort. Sci.* 53(3):298-302.

6) Fukuda, H., K. Chiba, and T. Kubota. 1980. Changes in texture properties of apple fruit by softening of the onset of mealiness. Bull. Fruit Tree Res. Sta. C. Japan. 10:33-47.

7) Fukuda, H., T. Kubota, and T. Suyama. 1979. Nondestructive measurement of chlorophyll, watercore, and internal browning disorders in apple fruits by light transmission. Bull. Fruit Tree Res. Sta. C. Japan. 6:27-54.

8) Fukuda, H. and T. Kubota. 1988. Nondestructive measurement of watercore in apple by light transmittance with single beam. Bull. Fruit Tree Res. Sta. (Morioka, Japan). 15:4147.

9) Fukumoto, M., K. Aoba, H. Yoshioka, and K. Nagai. 1988. Influence of the disorder bitter pit on microsomal vanadate-sensitive ATPase activity from apple fruit. *J. Japan Soc. Hort Sci.* 57(2):184-190.

10) Jones, J.M., J.B. Koen, S.A. Bound, and M.J. Oakford. 1991. Some reservations in thinning 'Fuji' apples with naphthalene acetic acids (NAA) and ethephon. *New Zealand Journal of Crop and Hort. Sci.* 19:224-228.

11. Kato, K., K. Abe, and R. Sato. 1977. The ripening of apple fruit. I. Changes in respiration, C_2H_4 evolution and internal C_2H_4 concentration during maturation and ripening. *J. Japan. Soc. Hort. Sci.* 46(3):380-388.

12) Kato, K., K. Goto, R. Sato, and R. Harada. 1978. The ripening of apple fruit. III. Changes in physiochemical components and quality, their interrelations and their relations to internal C_2H_4 concentration during maturation and ripening. *J. Japan. Soc. Hort. Sci.* 7(1):87-96.

13) Koganezawa, H., Y. Ohnuma, T. Sakuma, and H. Yanase. 1989. Apple fruit crinkle, a new graft-transmissible fruit disorder of apple. Bull. Fruit Tree Res. Stn. C. Japan. 16:57-62.

14) Michailides, T.J. and R.A. Spotts. 1990. Postharvest diseases of pome and stone fruits caused by *Mucor piriformis* in the Pacific Northwest and California. *Plant Dis.* 74(8):537-543.

15) Moore, J. 1991. Golden state apple boom. *Fruit Grower* 111(1):6-8.

16) Nakayama, M. and Y. Ota. 1983. Physiological function of ethylene and low pressure storage of apple. *JARQ* 17(1):30-33.

17) Okamoto, T., J. Harata, F. Nara, A. Osu, and K. Kobayashi. 1982. Influence of fruit bag treatment on the tree upon the keeping quality of apples during storage. Bull. Fac. Agric. Hirosaki Univ. Japan. 38:43-64.

18) Okamoto, T. and K. Iwasabi. 1985. Storage of apples in polyethylene containers. Bull. Agric. Hirosaki Univ. Japan 44:1-7.

19) Sadamori, S., et al. 1963. The new apple variety 'Fuji.' Bull. Hort. Res. Sta. C. (Morioka, Japan) No. 1:1-6.

20) Tsuchiya, S., Y. Yoskhida, T. Haniuda, and T. Sanada. 1975. Apple rootstock studies. II. Twelve years results on growth, cropping, and fruit quality with three varieties on M.9, *M. prunifolia*, and apple seedling rootstocks. Bull. Fruit Tree Res. Sta. C. Japan. 2:13-41.

21) Yanase, H., H. Koganezawa, and A. Yamaguchi. 1987. Occurrence of apple russet ring in Japan and its graft transmission. Bull. Fruit Tree Res. Sta. C. Japan. 14:53-60.

22) Yamazaki, T. and K. Suzuki. 1980. Color charts: useful guide to evaluation of fruit maturation. I. Colorimetric specification of color charts for Japanese pear, apple, peach, grape, kaki, and citrus fruits. Bull. Fruit Tree Res. Sta. A. Japan 7:19-44.

23) Yoshioka, H., K. Aoba, and M. Fukumoto. 1989. Relationships between qualitative and physiological changes during storage and maturation in apple fruit. *J. Japan Soc Hort. Sci.* 58(1):31-36.

24) Yoshida, Y., S. Tsuchiya, T. Haniuda, T. Sanada, and S. Sadamori. 1978. New apple variety 'Hatsuaki.' Bull. Fruit Tree Res. Sta. C. Japan 5:1-14.

Authors Yoshida and Fan were with the Estacao Experimental de Sao Joaquim, Santa Catarina, Brazil; and Patterson with the Department of Horticulture and Landscape Architecture, Washington State University, Pullman, Washington, when this article originally appeared, October 1995.

The 'Sumner' Pecan

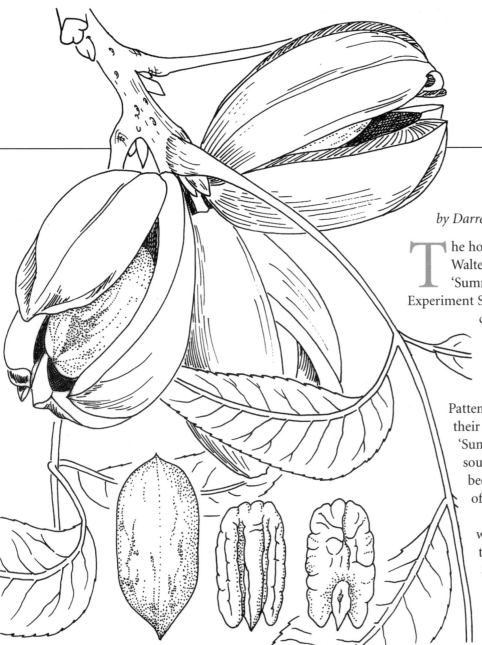

by Darrell Sparks

The horticultural value of this cultivar was recognized around 1932 by Walter E. Sumner, Tift County, Georgia (5). Eight years later, in 1940, 'Sumner' was included in the cultivar collection of the Coastal Plain Experiment Station, Tifton, Georgia. (12). Near the same time, 'Sumner' was commercially propagated by D.A. Law, Chula, Georgia, and by Fred Voigt, Waycross, Georgia. Voigt top-worked several trees to 'Sumner' and later propagated the cultivar in his nursery. Commercial planting was limited until Mr. Voigt sold 'Sumner' trees to Patten Seed Company, Lakeland, Georgia, in the late 1960s. Patten Seed Company established a pecan nursery with 'Sumner' as one of their major cultivars in 1977 and has been the major promoter of 'Sumner' since that time. Initially, 'Sumner' was planted primarily in the southeastern part of the state of Georgia; but now 'Sumner' is rapidly becoming a popular cultivar throughout the pecan producing region of Georgia.

The location of the original 'Sumner' tree was lost with time but was rediscovered by T.F. Crocker, D. Sparks, and I.E. Yates in 1993. The tree (Fig. 1) is among other pecan trees on the lawn of Phyllis S. Nicholson at the corner of 16th Street and Central Avenue, Tifton, Georgia. Most trees in the lawn have been top-worked to 'Sumner' indicating that a previous owner of the property recognized the value of this cultivar. One of the previous owners of the property was Thomas Daniel Smith, who planted nuts on the property in about 1905, according to his great granddaughter, Nellie Chapman Davis (personal communication, 1993). At least four seedlings came from this planting, two of which became the named cultivars, 'Big Z' and 'Sumner.' Many similarities between 'Sumner' and 'Schley' indicate that 'Sumner' may be a seedling of 'Schley.' Characteristics in common to the two cultivars include similar leaf form, oblong nut shape, consistent yearly nut production, and susceptibility to black pecan aphids.

Date of bud break is similar to 'Stuart' (9). Foliage is dark green and glossy. Foliage retention in the fall is excellent (7). Unlike most cultivars, mature trees of 'Sumner' rarely have a second cycle of shoot growth. Tree form is moderately upright (Fig. 2), and, as a result, 'Sumner' is more suited to high density planting than spreading cultivars such as 'Farley' and 'Desirable.' The tree develops strong crotches, requires very little training, and is resistant to breakage. The canopy is moderately open. Unlike 'Schley,' the bark of 'Sumner' is nonscaly (Fig. 3).

'Sumner' is protogynous. 'Sumner,' like 'Stuart,' is a good late-season pollinizer. 'Desirable' will pollinate 'Sumner' in Georgia (12).

'Sumner' is precocious and prolific (9). Fruit is borne on both short and long shoots from branches along the length of pole-like branches. Annual observations of mature trees in the Voigt orchard indicate alternate bearing is relatively minor compared to most cultivars and especially so for a prolific cultivar. The relatively consistent production is due in part to 'Sumner's' moderate second drop.

The shuck is thin (3), smooth, glossy, and has a distinguishing crease at the apex (Fig. 4). Date of nut maturity is very late, about 11 days after 'Stuart' (9). The late nut maturity is a major disadvantage.

Nut shape is oblong. The predominant base shape is obtuse, but base shape is variable and can vary to round and obtuse asymmetric. Apex shape is predominantly cuspidate, but can be cuspidate asymmetric. The apex is grooved along the sutures. Shell halves within a nut are usually equal. When shell halves are unequal, the smaller half is very much reduced and, in addition, is compressed in midsection. The overall appearance of these asymmetric nuts is boat-shaped, with the larger half forming the boat's bottom and sides. When the nut is symmetric, nut cross-section shape is round to slightly oval. The shell suture is elevated. Ridges are evident. Shell surface topography is smooth. The shell is dull, light brown with black markings. Both stripes and dots are moderate (9).

Nut shape is somewhat similar to 'Schley,' but, in contrast to 'Schley,' the nut is large at about 48 nuts per pound. On occasion, 'Sumner' nuts have been sold as "Jumbo Schley." The shell is thicker than 'Schley.' Percentage kernel is about 54. The nut continues to fill well as a mature tree. Kernel stability, unlike 'Schley,' is good. Color is acceptable with a color rating of 6.5 (1 = dark; 10 = light), but the kernel is darker in some years than in others. In addition, when left on the tree, the kernel darkens rapidly. Darkening can be

FIGURES 1 & 2.
Canopies and limb structure of 'Sumner.'

FIGURE 3.
The non-scaly bark of 'Sumner.'

FIGURE 4.
Fruit of 'Sumner' pecan has a smooth and glossy shuck with distinguishing crease at the apex.

minimized by early harvest (shortly after the shuck dehisces). However, when the wet nut is dried, color is sometimes uneven, in that part of the kernel is lighter than the remainder of the kernel. Although marginal kernel quality is often cited as a disadvantage, the criticism does not appear to be valid, as the kernel sells well in the shelled trade. The kernel has deep, narrow dorsal grooves, which makes removal of the packing tissue somewhat difficult. In addition, the kernel has a tendency to chip during the cracking process. Regardless, the nut brings a premium price in the in-shell trade. However, the narrow grooves increase processing costs, and, for this reason, the nut is sometimes downgraded by the processor (9), especially in years when the national pecan production is high.

'Sumner' has good resistance or immunity to scab (8). During the excessively wet growing season of 1991, 'Sumner' was immune to scab in some locations in Georgia, but mild scab has been observed on 'Sumner' in other locations. As in 'Schley,' 'Sumner' foliage is especially susceptible to early season black pecan aphid infestations which are also a problem in late season. The leaf has moderate resistance to sooty mold accumulation (9). 'Sumner' is medium in susceptibility to hickory shuckworm (3) and nut curculio, but has above-average resistance to pecan bud moth (6). At Brownwood, Texas, the leaf is immune to southern pecan leaf phylloxera (1) and has high resistance to pecan phylloxera (2). The fruit is highly susceptible to damage from stink bug. 'Sumner' nursery trees are very susceptible to both winter (11) and early fall freezes (4). Freeze damage can be greatly minmized by propagation to a seedling (juvenile) trunk (10). In some seasons, poor shuck opening and premature germination of nuts on the tree can be a problem. 'Sumner' fruits are subject to splitting during the water stage of development. The split is due to excessive internal pressure that occurs with high humidity and soil moisture as in 'Wichita' (9). 'Sumner' is easy to propagate.

Scab immunity or high resistance, good nut size with a well-developed kernel, heavy nut production at a young tree age, and the tendency to produce relatively consistently on an annual basis have made 'Sumner' a popular cultivar for new plantings by many growers in Georgia. The popularity of this cultivar has increased rapidly in spite of late nut maturity and less than desirable cracking ability.

Literature cited

1) Calcote, V.R. 1983. Southern pecan leaf phylloxera *(Homoptera:Phylloxeridae)*: Clonal resistance and technique for evaluation. *Environ. Entomol.* 12:916-918.

2) Calcote, V.R. and D.E. Hyder. 1980. Pecan cultivars tested for resistance to pecan phylloxera. *J. Georgia Entomol. Soc.* 15:428-431.

3) Calcote, V.R., G.D. Madden, and H.D. Peterson. 1977. Pecan cultivars tested for resistance to hickory shuckworm. *Pecan Quarterly* 11(1):4-5.

4) Goff, W.D. and T.W. Tyson. 1991. Fall freeze damage to 30 genotypes of young pecan trees. *Fruit Var. J.* 45:176-179.

5) Leidner, J. 1982. Sumner: An old pecan variety that's proving itself. *Prog. Farmer* 97(4):52.

6) Mizel, R.F. and D.E. Schiffhauer. 1986. Larval infestation levels of pecan bud moth, *Gretchena bollinana* (Lepidoptera: Tortricidae), in relation to cultivar and position on the tree. *Environ. Entomol.* 15:436-438.

7) O'Barr, R.D. and S. Rachal. 1987. Louisiana pecan variety update. *Pecan South* 21(4):12-16, 27, 28.

8) Sanderlin, R.S. 1987. Evaluation of pecan cultivars and USDA selections for scab disease susceptibility. Res. Rpt. Pecan Res.Ext. Sta., Louisiana Agr. Exp. Sta., Shreveport, Lousiana.

9) Sparks, D. 1992. Pecan cultivars—The orchard's foundation. *Pecan Production Innovations,* Watkinsville, Georgia.

10) Sparks, D. and J.A. Payne. 1977. Freeze injury susceptibility of non-juvenile trunks in pecan. *HortScience* 12:497-498.

11) Sparks, D. and J.A. Payne. 1978. Winter injury in pecans—A review. *Pecan South* 5(2):56-60, 82-88.

12) Worley, R.E., and O.J. Woodard, and B. Mullinix. 1983. Pecan cultivar performance at the Coastal Plain Experiment Station. Univ. Georgia Agr. Exp. Sta. Res. Bull. 295.

Author Sparks was with the Department of Horticulture, University of Georgia, Athens, Georgia, when this article originally appeared, January 1996.

The 'Crandall' Black Currant

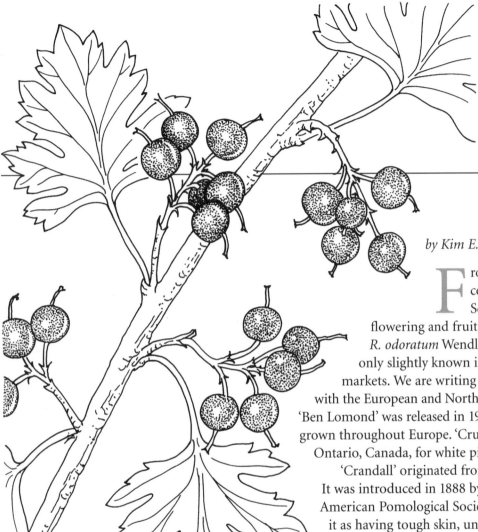

by Kim E. Hummer & Stanislaw Pluta

From May to August 1995, 116 *Ribes* cultivars and selections in the collection at the U.S. Department of Agriculture, Agriculture Research Service, National Clonal Germplasm Repository, were evaluated for flowering and fruiting characteristics. One black-fruited, late-ripening currant cultivar *R. odoratum* Wendl. cv. Crandall was exceptional for berry size and quality. 'Crandall' is only slightly known in the United States, and is unknown to European gardens and markets. We are writing this article to emphasize some of 'Crandall's' qualities in contrast with the European and North American black currants *R. nigrum* cvs. 'Ben Lomond' and 'Crusader.' 'Ben Lomond' was released in 1975 by the Scottish Crop Research Institute and is a prominent cultivar grown throughout Europe. 'Crusader' was released in 1948 from the Department of Agriculture, Ontario, Canada, for white pine blister rust resistance.

'Crandall' originated from a wild seedling discovered by R.W. Crandall of Newton, Kansas. It was introduced in 1888 by Frank Ford & Sons, Ravenna, Ohio, and recommended by the American Pomological Society in 1899 (2). 'Crandall' has a poor reputation. Hedrick (2) describes it as having tough skin, unpleasant flavor, and uneven ripening. Our assessment of 'Crandall' is in sharp contrast. The berry skin of 'Crandall' does not seem tougher than that of *R. nigrum* L. cultivars, and the flavor is sweet and pleasant. The fruit ripens consistently enough on the plants in the Repository collection but may be too droopy for mechanical harvesting.

Ribes odoratum, the species from which 'Crandall' was selected, is taxonomically placed in a section called Symphocalyx, the golden currants. A second species of this section is *R. aureum* Pursh. Both *R. odoratum* and *R. aureum* have bright yellow fragrant flowers with a tubular receptacle which ranges up to 10 mm. The petals are usually red-tipped in mature flowers. The shrubs are erect to 2.5 meters high with many suckers growing from the crown. While *R. odoratum* is native east of the Rocky Mountains, *R. aureum* is native to the West (3). These species are cultivated throughout North America as ornamental flowering shrubs. Although the shrub has a somewhat undesirable, stiff, upright growth habit during the summer, the yellow flowers which bloom in mid-April, and the yellow and red fall foliage provide added landscaping value. *Ribes odoratum* is used as a rootstock for grafting other cultivated currants and gooseberries (1).

Under Corvallis, Oregon, conditions, 'Crandall' is a vigorous plant growing more than 2 m in a season, which is slightly more than 'Ben Lomond' and much more than 'Crusader.' In cooler, drier environments, 'Crandall' may grow ≤1.5 m per season. 'Crandall' fruits have a smooth bluish-black skin, are about twice as large as those of 'Ben Lomond, and three times as large as those of 'Crusader' (Table 1). The calyx is persistent in each of the three cultivars. 'Crandall' flesh is whitish-yellow and not very juicy, while the *R. nigrum* berries have juicy, darker flesh. 'Crandall' berries are sweeter, and the seeds are not as noticeable as those of *R. nigrum* cultivars. 'Crandall' has typically less than 10 seeds per berry, while *R. nigrum* cultivars have 10 to 30. 'Crandall' fruits are born in loose, short clusters with 2 to 6 berries per inflorescence, i.e., strig, whereas 'Ben Lomond' has from 4 to 8, and 'Crusader' has 5 to 10. 'Crandall' does not have the heavy odor of *R. nigrum* cultivars.

'Crandall' plants are resistant to white pine blister rust, *Cronartium ribicola* Fisher, leaf spot,

TABLE 1.

Flower and fruit descriptors for R. odoratum Wendl, cv. 'Crandall' as compared to R. nigrum L. cvs. 'Ben Lomond' and 'Crusader.'

Cultivar	Flower season	Plant habit^Z	Vigor^Y	Harvest date	Yield^X	Fruit size g/berry^W
Crandall	midseason	4	9	very late, July 28	6	2.17c
Crusader	early	8	7	early, June 29	4	0.69a
Ben Lomond	late	7	8	late, July 19	6	1.36b

^{Z} Plant habit was rated on a scale from 1, spreading, to 9, very erect.
^{Y} Vigor was rated on a scale from 1, weak, to 9, very green, healthy, rapidly growing plants.
^{X} Yield was rated on a scale from 1, poor, to 9, excellent production.
^{W} These figures display the average weight of 3 replicates of 100 berries. Means were separated using LSD at $p < 0.01$.

TABLE 2.

Disease effects on R. odoratum Wendl, cv. 'Crandall' and R. nigrum L. cvs. 'Ben Lomond' and 'Crusader' in Corvallis, Oregon, in 1995. Spring frost/Botrytis was evaluated May 10-15; powdery mildew, August 8; leaf spot, August 15; white pine blister rust, September 18.

Cultivar	Spring frost/Botrytis^Z	Powdery mildew	Leaf spot	White pine blister rust
'Crandall'	5	1	1	1
'Crusader'	2	3	4	1
'Ben Lomond'	2	2	2	5

^{Z} These diseases were rated on a scale of 1, no symptoms, to 9, severe symptoms.

This late-ripening cultivar could extend the European black currant production season.

Authors Hummer and Pluta were with the USDA-ARS NCGR, Corvallis, Oregon; and the Institute of Pomology and Floriculture, Skierniewice, Poland, respectively, when this article originally appeared, April 1996.

Drepanopeziza ribis (Kleb.) Hohn, and American powdery mildew, *Sphaerotheca mors-uvae* (Schwein.) Berk. & Curt., even during years of heavy infestation on other cultivars (Table 2). In Poland, 'Ben Lomond' shows a rating of 4 to 5 for powdery mildew and white pine blister rust in most years (Pluta, unpublished).

Unfortunately, spring frosts can damage 'Crandall' blossoms and stems during bloom, after which gray mold, *Botrytis cinerea* Pers., can enter and cause stem dieback (Table 2). Both *R. aureurn* and *R. odoratum*, including 'Crandall,' are susceptible to spring frost and *Botrytis* damage. During June and July 1995, 'Crandall' suckers grew rapidly to replace canes damaged by spring frost and *Botrytis*. The other two black currants were not injured as severely (Table 2).

Because the fruits are so large, and the plant is vigorous and healthy, 'Crandall' would be a good cultivar for landscaping, backyard gardening, or small commercial operations such as pick-your-own. This late-ripening cultivar could extend the European black currant production season. Hedrick (2) mentions that 'Crandall' survives in regions such as the midwest of the United States, which have hot dry summers.

Some European markets are expanding the production of black currants for fresh sales. 'Crandall' may have direct marketing value for fresh production, or for processing, as in jams or jelly. This cultivar will be examined for breeding potential in the Polish *Ribes* program.

Literature cited

1) Galletta, G. and D. Himelrick. 1990. *Small fruit crop management.* Prentice Hall, Englewood Cliffs, NJ. 602 pp.

2) Hedrick, U.P. 1925. *The small fruits of New York.* J.B. Lyon. Albany, NY.

3) Hitchcock, C.L. and A. Cronquist. 1984. Part 3: Saxifragaceae to Ericaceae. In: *Vascular Plants of the Pacific Northwest.* C.L. Hitchcock et al., eds. University of Washington Press. Seattle.

The 'Kent' Strawberry

by Andrew R. Jamieson

The 'Kent' strawberry, since its introduction in 1981, has rapidly risen to become the most widely planted strawberry cultivar in Canada. It replaced 'Redcoat' as the leading midseason strawberry of eastern Canada and the prairie provinces. Compared to 'Redcoat,' 'Kent' was originally described as a more productive and larger fruited strawberry (4). Strawberry growers quickly confirmed these claims, and by 1985, 'Kent' was the leading cultivar planted in Nova Scotia and soon thereafter became important for growers in Quebec, Ontario, and Manitoba (5). 'Kent' did not stand up to the prevailing complex of viruses in British Columbia, where 'Totem' predominates (4).

'Kent' was selected in 1974 at the Kentville Research Station of Agriculture Canada from 198 seedlings of ('Redgauntlet' x 'Tioga') x 'Raritan.' 'Kent' was the fourth strawberry introduced by D.L. Craig in the breeding program he began in 1949. The broad adaptability of 'Kent' is remarkable. Its high productivity has been demonstrated in Atlantic Canada (4, 7), Ontario (6), Minnesota (16), Missouri (14), and Colorado (18). In 1988, it was listed as promising for production in central and eastern Canada, and in the north central, mid central, northeastern, and middle Atlantic regions of the United States (9, 12).

By 1990, 'Kent' had become an important cultivar not only in Eastern Canada, but also in the Upper Midwest and Northeastern USA (2). 'Tioga' and 'Redgauntlet,' both grandparents of 'Kent,' may have contributed to its broad adaptability. Daubeny has stated that, in many respects, these two cultivars have the broadest adaptations of all cultivars (10).

> *"A pleasant but undistinguished flavor is regrettably the answer—following the maxim that most will like that which has nothing to dislike."*
>
> A.G. Brown, on apple flavor, which, the author states, may be descriptive of 'Kent's' flavor.

One contributing factor to high yields is the extended harvest period of 'Kent' due to the good fertility of late series flowers (3). Another factor is that fruit size does not decline as fast as with many other varieties, such as 'Blomidon' or 'Bounty.' In small research plots, 'Kent' has yielded 40 tonnes per hectare but 25 to 30 tonnes per hectare is more common (4, 13). In a commercial setting using matted row culture, excellent yields are produced in the first and second cropping years but this performance may not be achieved in later years.

The fruit of 'Kent' are as large or larger than most commercially grown cultivars in the Northeast, with the exception of 'Cavendish,' and they are globose-conic, firm, attractive, and glossy (13, 17). Fruit color is typically bright red but can become dark under excessively hot, ripening weather. Under such conditions, the skin is often weak, and this may limit its usefulness in the south-central part of its productive range.

The flavor of 'Kent' is relatively weak, compared with 'Honeoye' and 'Cavendish' (13). When grown at Kentville, 'Kent' was perceived as less tart than 'Honeoye' with correspondingly lower titratable acidity, which contributed to a higher overall acceptance by sensory panelists (13). Aroma volatiles continue to develop after harvest at a greater rate in 'Kent' than 'Honeoye' (Forney, pers. comm.). The comments of A.G. Brown on apple flavor, "A pleasant but undistinguished flavor is regrettably the answer—following the maxim that most will like that which has nothing to dislike" is appropriate for 'Kent' (1). 'Kent' has been rated 'good' as a frozen sugar-packed product (19).

Plants of 'Kent' are vigorous, and runner production is usually sufficient to fill a matted row but runnering is not excessive. Plants are winter-hardy resistant to powdery mildew and moderately resistant to leaf scorch but susceptible to leaf spot, *Verticillium* wilt, red stele root rot, gray mold, and anthracnose fruit rot (15, 17). Leaves are resistant to two-spotted spider mites, as measured by a leaf disk bioassay (8). The cultivar is attractive to and injured by tarnished plant bugs (11). Plants are tolerant to several herbicides registered for use on strawberries, but they are sensitive to terbacil.

A challenge for strawberry breeders is to develop cultivars which match 'Kent' in yield but surpass it in fruit size, fruit quality, and disease resistance. Important aspects of fruit quality to be improved are flavor intensity, skin toughness, and quality retention during shipping. In terms of diseases, resistance to the many races of *Phytophthora fragariae* (Hickman), the cause of red stele root rot, would be a great asset in the Northeast. Unfortunately, 'Kent' has not proven to be a particularly useful parent, despite being frequently included in breeding programs. However, it is likely that strawberry growers will find 'Kent' a profitable midseason cultivar for several years to come.

In recognition of his origination of several commercially important berry cultivars, Donald Craig was the 1990 Wilder Medal recipient. 'Kent' is unquestionably his greatest success.

Literature cited

1) Brown, A.G. 1975. Apples. pp. 3-37. *In:* J. Janick and J.N. Moore (eds.) *Advances in Fruit Breeding.* Purdue University Press, West Lafayette, IN.

2) Chandler, C.K. 1991. North American Strawberry Cultivars. pp. 60-65. *In:* A. Dale and J.J. Luby (eds.), *The Strawberry into the 21st Century.* Timber Press, Portland, OR.

3) Chercuitte, L., J.A. Sullivan, Y.D. Desjardins, and R. Dedard. 1991. Yield potential and vegetative growth of summer-planted strawberry. *J. Amer. Soc. Hort. Sci.* 116:930-936.

4) Craig, D.L., L.E. Aalders, and G.W. Bishop. 1982. Kent strawberry. *Can. J. Plant Sci.* 62:819-822.

5) Dale, A. 1989. Eastern Canada strawberry cultivars. *Fruit Var. J.* 43:38-41.

6) Dale, A. (ed). 1989. Ontario Coordinated Berry Crop Cultivar Trials. Horticultural Research Institute of Ontario, Simcoe, Ontario.

7) Estabrooks, E.N. and M. Luffman. 1989. Strawberry cultivar evaluation in New Brunswick, Canada. Adv. Strawberry Prod. 8:58-61.

8) Gimenez-Ferrer, R.M., J.C. Scheerens, and W.A. Erb. 1993. In vitro screening of 76 strawberry cultivars for two-spotted spider mite resistance. *HortScience* 28:841-844.

9) Hancock, J.F. and D.H. Scott. 1988. Strawberry cultivars and worldwide patterns of strawberry production. *Fruit Var. J.* 42:102-108.

10) Hancock, J.F., J.L. Maas, C.H. Shanks, P.J. Breen, and J.J. Luby. 1991. Strawberries *(Fragaria)*. pp. 491-546. *In:* J.N. Moore and J.R. Ballington (eds.), "Genetic Resources of Temperate Fruit and Nut Crops." *Acta Hort.* 290, Wageningen:ISHS.

11) Handley, D.T., J.E. Dill, and J.E. Pollard. 1993. Tarnished plant bug injury on six strawberry cultivars treated with differing numbers of insecticide sprays. *Fruit Var. J.* 47:133-137.

12) Hansen, E.J. 1989. Performance of strawberry cultivars in the north central region of the United States. *Fruit Var. J.* 43:151-154.

13) Jamieson, A.R., K.A. Sanford, and N.L. Nickerson. 1991. 'Cavendish' strawberry. *HortScience* 26:1561-1563.

14) Kaps, M.L., M.B. Odneal, J.F. Moore, and R.E. Carter. 1990. Strawberry cultivar evaluation in Missouri. *Fruit Var. J.* 74:158-164.

15) LaMondia, J.A. 1991. First report of strawberry anthracnose caused by *Colletotrichum acutatum* in Connecticut. *Plant Disease* 75:1286.

16) Luby, J.J., E.E. Hoover, S.T. Munson, D.S. Bedford, D.K. Wildung, and W. Gray. 1984. Performance of strawberry cultivars in Minnesota: 1983. *Adv. Strawberry Prod.* 3:11-14.

17) Luby, J.J. 1989. Midwest and plains states strawberry cultivars. *Fruit Var. J.* 43:22-31.

18) Renquist, A.R. and H.G. Hughes. 1992. 'Kent' and 'Honeoye' were highest yielding, best adapted strawberry cultivars in Colorado trial. *Fruit Var. J.* 46:58-61.

19) Wang, S.L. and A. Dale. Evaluation of strawberry cultivars for frozen sugar pack. *Adv. Strawberry Prod.* 9:31-32.

Author Jamieson was with Agriculture and Agri-Food Canada, Kentville Research Centre, Kentville, Nova Scotia, Canada, when this article originally appeared, July 1996.

French Hybrid Grapes in North America

by G.A. Cahoon

By the simplest definition, French hybrids are a group of grape selections that originated in France from crossing European *(Vitis vinifera)* varieties with certain North American species. Selections were from both wild types and early nineteenth century cultivars. The name French-American hybrids is also frequently used in the United States to identify them. To further amplify how ambiguous the hybrid grape picture has become, newer varieties with French hybrid parentage have also been developed in the United States and Canada. These are generally categorized as American hybrids, or American varieties.

The French hybrids have made their home here in the United States and Canada mostly east of the Rocky Mountains, and appear to be serving a well-defined role for the North American wine industry. The pioneer French breeders were Eugene Contassot, Albert Seibel, and Georges Couderc. Other early contributors Fernand Gaillard, Francois and Maurice Baco, Bertille Seyve, Eugene Kuhlmann, Pierre Castel. Christian Oberlin, and others, primarily used the species *V. rupestris, V. lincecumii* and *V. riparia* in their crosses.

By the time World War II came, a new group of French hybridizers, Bertille Seyve-Villard, Joannes Seyve, J-F Ravat, Joanny Burdin, Jean-Louis Vidal, Alfred Galibert, Pierre Landot, Eugene Rudelin, and others, were playing an important role. Collectively, they had been intensively carrying on this work of hybridization for three quarters of a century when individuals in the United States became interested in them during the late 1930s and early 1940s. In 1948, O.A. Bradt from the Horticultural Research Institute of Ontario, Vineland Station, stated, "In recent years, there has been an increasing interest in the French hybrid grapes, particularly from the wine standpoint" (4, 5). This statement is still appropriate in the 1990s, as the hybrids fill a significant role in many, if not most, eastern and midwestern wineries, both as varietals and for blending. However, the projections for the future may not be as bright as in previous years. The mood of many eastern wineries is shifting toward the traditional *V. vinifera* cultivars.

Phylloxera

Since the grape Phylloxera, *Daktulosphaira vitifolie* (Fitch), the grape root louse, was so important in the creation of the French hybrids, it is deemed appropriate that some early history be provided.

In the Bushberg Catalog, 3rd edition 1883 (23), C.V. Riley stated: "Among the insects injurious to the grape-vine, none have attracted as much attention as the phylloxera, which, in its essential characteristics, was unknown when the first edition of this little-known work on "American Grape Vines" was written (1869). This gall-forming insect, it is true, was noticed by North America grape-growers many years ago (especially on the 'Clinton'), but they knew nothing of its root-inhabiting form. Even Fuller—who informs us that in Mr. Grant's celebrated grape nurseries (as far back as 1838), the men were in the habit of 'combing out' with their fingers, the roots of young vines to be sent off, in order to get rid of 'knots'—never mentioned anything of this, nor of any root-infesting insect, in his excellent "Treatise on the Cultivation of the Native Grape," though 16 pages are devoted to grape insects.

He also stated that in 1869, M.J. Lichtenstein, of Montpellier, France, first hazarded the opinion that phylloxera, which was attracting so much attention in Europe, was identical with the American leaf-gall louse (first described by Dr. Asa Fitch, state entomologist of New York, by the name of *Pemphigus vitifoliae*. In 1870, Prof. C.V. Riley succeeded in establishing the identity of the gall insect with those from Europe and also the identity of the gall and root-inhabiting types. This was later confirmed by other scientists in France and Austria.

Professor Planchon from Montpellier also visited the United States and confirmed these observations. Later, Bushberg Nursery sent "a few thousand cuttings," gratis, to Montpellier, and the success of these tests has resulted in an immense demand for the resistant varieties." Together, Profs. Riley and Planchon established that the insect was indigenous to the North American continent, east of the Rocky Mountains, and was imported into Europe on American vines. Later, as this information became public knowledge, government regulations were set up to exclude not just the insect, but the remedy: wood from the resistant American rootstock varieties onto which the susceptible European *V. vinifera* varieties could be grafted.

Eventually, it was the grafting of traditional varieties onto resistant rootstock varieties developed by such breeders as Millardet, de Grasset, Couderc, and Ganzin, that allowed the resurgence of the European *V. vinifera* grape industry. These rootstock hybrids used the American species *V. riparia*, *V. rupestris*, and *V. berlandieri*.

A need for new varieties

The need for phylloxera-resistant grape varieties was not the only reason French hybrids came into existence. In 1847, powdery mildew (oidium) (*Uncinula necator* Burr) was introduced into French vineyards from North America through botanists and other plant collectors. There is also evidence that downy mildew (*Plasmopara viticola* Berl.) and black rot (*Guignardia bidwellii* Ellis) arrived about 1878. Some chemical treatments had been discovered (Bordeaux mix and sulfur), but they required special equipment and were costly. Thus, a multiple need for resistant varieties existed: to avoid extermination from phylloxera, diseases, and to hold down production costs.

It had been observed in France that some of the wild American grape species and a few American varieties exhibited some resistance, or at least did not succumb as rapidly to these pests as did *V. vinifera* varieties. American varieties with *V. labrusca* parentage were deemed undesirable due to their objectionable, "foxy" flavor. But other wild species did not have this objectionable characteristic and had some resistance to phylloxera and some diseases. Thus, large quantities of wild American grape seeds were shipped to or collected by French viticulturists, and the work to create new varieties began.

Hybrid breeding programs

Following the initial efforts of the rootstock breeders, as early as 1874 (17), programs were active at Montpellier to produce hybrids resistant to phylloxera, diseases, and eventually to calcarious soils, and yet capable of producing acceptable fruit for wine on their own roots. Collectively, they became known as *"Les hybrides producteurs directes"* (the hybrid direct producers, HPDs).

Some of the breeders, such as Foex and Ravaz, were always cautious or opposed to the propagation of hybrids in quality vineyards, but were enthusiastic about the rootstocks (17). According to the records (2, 13, 16, 17, 25) and contrary to public opinion, the first HPDs came from the United States: 'Clinton,' 'Isabelle,' 'Othello,' 'Noah,' 'Herbemont,' 'Jacquez,' and others. For the most part, they were hybrids of *V. labrusca, V. aestivalis, V. riparia,* and sometimes *V. vinifera*. The taste of these foreign varieties caused them to be generally unacceptable in France. However, because of their productivity and disease resistance, several of them were extensively planted.

By contrast, the breeding programs conducted by previously mentioned French breeders, with *V. vinifera* as the major parent crossed primarily with *V. rupestris* and *V. lincecumii*, developed a group of HPD selections, and is the focus of our present discussion. Varieties with *V. labrusca* parentage were seldom used by the French.

The HPDs acquired good acceptance by French growers, and considerable acreages were eventually planted. According to Wagner (29), by 1955, 35% of the table wine produced was of hybrid origin. In 1958, Galet (17) stated this amount as 42%. None, however, had been granted the right of "appellation d'origine." Hybrids had been forbidden in the Appelation d'Origine Contrôlée (AOC) and Vin de Qualité Supérieure (VDQS) since 1927. Their use was for ordinary wines with no appellation (vin ordiniare) and table grapes. It is also worthy of note that the goal of the later breeders was to back-cross the best of the hybrids with the best of the *V. vinifera* varieties rather than cross the best of the hybrids with the best of the hybrids.

During this latter period and up to the present day, there have been restrictions on the planting and sale of HPDs. Earliest restrictions were by the French in 1953, followed by restrictions in the Common Market in 1976 (16, 17). Restrictions by the French have always been more severe than those of the Common Market. As a result, acreage has significantly decreased. However, in 1900, there were 123,500 acres (50,000 ha) of hybrids in France. By 1929, plantings had grown to 534,007 acres (216,197 ha) (14.5%) and to 1,005,368 acres (402,147 ha) by 1958 (31%) (16, 17); in 1968, there were 726,904 acres (294,293 ha) (23.5%); and in 1979, less than 179,102 acres (72,511 ha) (14%). Wagner, in 1978 (29), stated that the 13 most popular hybrids in France occupied 487,000 acres in 1974, an acreage greater than all the true wine grapes in California at the time. Wagner (29), quoting from Galet's work, indicated that all the 'Cabernet Sauvignon' in France around Bordeaux and other regions totaled 24,000 acres. Comparatively, this was the same as the acreage of 'Baco noir.' In 1990, the acreage of HPDs in France was estimated to be 87,000 acres (3.5%) (17).

Where did it start in the United States?

A search into where the impetus came for the exploration and use of French hybrids in the United States most logically leads to Philip and Jocelyn Wagner of Boordy Vineyards, Riderwood, Maryland, a suburb of Baltimore (Fig. 1). In the spring of 1935, as an amateur wine maker, Philip Wagner had planted some grafted vinifera: 'Carignane,' 'Sirah,' 'Franken Riesling,' 'Semillon,' and others. All, with the exception of 'Chardonnay,' failed.

The Wagners first became involved with the French hybrids in the late 1930s when Philip was working as London correspondent for the *Baltimore Evening Sun*. In a British publication about Greece, various aspects of restoring Greek agriculture were discussed. Pertaining to viticulture, it recommended a series of grape varieties, including several French hybrids. He also visited East Malling, where researchers were studying the French hybrids. As editor of the *Baltimore Sun*, he traveled back and forth to France, and, in his search of the French literature, came upon references to the grapes known as HPDs: the hybrid direct producers. Visits to the New York Agricultural Experiment Station, Geneva and Fredonia, New York, also opened up other options, and trials of some of these materials were begun.

Some of their early experiences with HPDs were had while traveling through the wine districts of France, where they saw hybrids being grown and harvested in the Dijon and Beaune area. During these travels, they also became acquainted with the individuals breeding the hybrids, and a very articulate person by the name of Gerard Marot, the president of "FENAVINO" (Fédération Nationale de la Viticulture Nouvelle et du Syndicat des Hybrideurs et Métisseurs Viticoles). The official monthly publication

> *"Because once one got interested in them, the numbers had a music all their own."*
>
> Philip Wagner, of the numbering system for French hybrid wine grape varieties.

FIGURE 1.
Philip and Jocelyn Wagner, owners and operators of Boordy Vineyards and pioneers in the introduction and expansion of French hybrid grapes in the United States (1992).

devoted to the defense of the French hybrids was the *La Viticulture Nouvelle*. Many misconceptions concerning the hybrids were widespread in France and soon found fertile ground in the United States. The HPD breeders were a very modest group and had to combat powerful anti-hybrid forces. They were not looked on favorably by the scientific community but as a group that just crossed one grape with another (28). The big industrial producers were the anti-hybrid people of southern France. Wherever the hybrids grew, the vinifera producers would lose a part of their market. These people had heavy influence with the viticultural establishment and a big block of members in the Chamber of Deputies. They also dominated the press and media. Most of the HPD breeders were so circumscribed by this time that they found it difficult to exchange promising hybrids with friends or associates.

Introductions and trials in the United States and Canada

The Wagners were not among the very first people to introduce French hybrids (HPDs) into the United States. But, since their first explorations into the feasibility of growing the hybrids in 1936-37, they have probably done more to further their cultivation and progress than any other individuals. The French hybrids, they found, already existed in obscure collections of several individuals and organizations. George Husmann had some hybrids shipped into California in 1911, and Hilgard may have brought some into the United States before the turn of the century.

They were not being studied or utilized at this time. Hybrids also existed at the New York Agricultural Experiment Station as early as 1911, and may have been from the group introduced by Husmann (9). Active in their introduction and cultivation were Fred Gladwin at Fredonia and Richard Wellington at Geneva.

The U.S. Department of Agriculture (USDA) Field Station at Fresno, California, obtained some hybrids during the late 1920s and early 1930s. A chapter in the 1937 *Yearbook of Agriculture* by Elmer Snyder (26), contains a very comprehensive review of grape breeding in the United States and lists the French hybrids in their Oakville collection. The University of California at Davis later acquired a substantial number of the hybrid selections in their collection. Two shipments of 20 selections each were introduced from France in 1927 and 1939 by the New York Agricultural Experiment Station at Geneva (9). Among the first shipment were 'Seibel 1000' and 'Seibel 5898.' 'Seibel 7053' was included in the second group.

To Philip Wagner, this early period was a very exciting experience, and his notes after viewing the collections of R.T. Dunstan (12) and J.R. Brooks in North Carolina stated: "Their collections are a sufficient basis for an entirely new viticulture" (29).

The Wagners became very active with the hybrids and facilitated the importation and exchange of materials with many individuals and organizations in the eastern United States and Canada. Through this period, Wagner's Boordy Vineyards nursery, established in 1941, represented the only source of rooted hybrids. It was not until 1952 that the New York Fruit Testing Association, listed its first HPDs. But through this entire period, and even today, the nursery at Boordy Vineyards is a source of hybrid vines. A large acreage of the plantings along the whole Chatauqua grape belt, for Taylor's winery and several others, was provided by Foster Nurseries, Fredonia, New York. Owner Walter Foster Gloor, was eventually succeeded by his son Bob Gloor, who carried on the business until 1988. In Ontario, Mori's Nursery and several wineries produced vines for many new hybrid vineyards. Large quantities of wood were also sent to British Columbia nurseries for propagation.

To get possession of hybrids, the Wagners picked them up everywhere: France; the New York Agricultural Experiment Station at Fredonia and Geneva, New York; USDA Field Station, Fresno, California; R.T. Dunstan, Greensboro, North Carolina; Napa Valley, Canada; and other places. Dunstan's sources were from quite an independent starting point from the others. He obtained a series of HPDs of Seibel and Seyve-Villard from France in 1938. Subsequent selections were collected from the hybridizers Galibert, Perbos, Joannes Seyve, Vidal, Landot, and Couderc. S.V.12-375 could have come from this source.

George Hostetter, Brights Wines Ltd., Ontario, Canada, imported some hybrids in 1943. 'Seibel 1087' was grown and wine marketed by them in 1946-47 (5). However, Jordan Winery is acknowledged as the first to market wine in quantity from 'Seibel 9549.'

In the Finger Lakes, according to Alexander Brailow and Philip Wagner (7, 28), one of the earliest commercial French hybrid vineyards was planted in 1944 by Gold Seal on the east side of Keuka Lake. It came about when Ed Sillsman, a Frenchman and a salesman for Gold Seal, had dinner at the Wagners' and a bottle of 'Baco #1,' red wine, was served. Sillsman was delighted at the quality (28). Gold Seal ordered 'Seibel 1000' vines and planted the vineyard. The vines were described in later years as having "trunks as big as oak trees." The vineyard may now have been replaced, but it existed for 30 to 40 years.

When eastern American wineries began to use the hybrids, it was only for blending. Many eventually were named because the winery people thought it would be impossible to sell a wine with a number such as 'Seibel 5279,' 'Seibel 9549,' etc. Philip Wagner (28) didn't find the numbers offensive. In fact, the names didn't mean very much to him. "Because once one got interested in them, the numbers had a music all their own."

There were periodic meetings between wineries and research people to evaluate progress with the hybrids and communicate their findings. At one such meeting at Fredonia, in the fall of 1945, Alex Brailow (7), winemaker at Gold Seal, recalls the following: "We had a whole lot of samples, and all we had was a bucket and jelly glasses." Richard Wellington was very important and very influential in the discussions that went on at this time with the French hybrids.

Adhemar de Chaunac brought some of the best Canadian-made wines to such a meeting. He was especially proud of the port wine made from the variety President; also included were some wines from Munson varieties. When they started tasting, the hybrids showed up so well that Adhemar could hardly believe it. It should be noted that about this time the Canadians were especially worried about wine imports from such countries as South Africa and Australia, because the Commonwealth had not provided any trade barriers of any sort. Their immediate concern was that they had to grow better grapes. The French hybrids now appeared to be at least a partial solution to these problems. Such selections as 'Seibel 1000' and 'Seibel 2008,' probably obtained from U.S. sources, had been planted in Ontario in 1918 and 1943.

Following the New York meeting, several actions were initiated to introduce more hybrids into Ontario. Jake Hostetter, foreman at Brights Wines at this time, was contacted by the vice president and asked to help the Ontario Ministry of Agriculture (OMAC) obtain vines or cutting of the hybrids (14).

In 1946, two shipments of hybrids were obtained from France—one for Brights and one for the Horticultural Experiment Station. The number of vines obtained by Brights was much smaller than the one for the OMAC but consisted of many selections (14). Contacts were made with French nurseries to import the hybrids. Records are not available as to what was obtained for Brights.

The shipment for the Ontario Ministry of Agriculture was received in late 1946 by the Horticultural Experiment Station, Vineland Station, Ontario, after many bureaucratic roadblocks and six weeks at sea (14). Included were 39 numbered selections as cuttings, rooted vines, and grafted vines. Richard Wellington, as well as Adhemar de Chaunac and Jake Hostetter, were influential in helping find nursery stock. O.A. Bradt (5) remembers they were in bad shape on arrival. One row of each (150 vines) was planted into the vineyard at the Rittenhouse Grape Substation in the fall of 1947 and the spring of 1948—half on sandy soil and half on heavy clay soil. 'Seibel 9549' came in as cuttings in late spring of 1947 and was planted out in the fall of 1947 and spring of 1948. Materials were obtained from Seibel, Bertille-Seyve, and Seyve Villard.

Of the Taylor brothers, Greyton was the one most interested in making wine from the hybrids. The photos (Figs. 2 and 3) are from meetings at Taylor's and Pleasant Valley Wine companies in 1951 and 1952. Richard Wellington, John Einset, and Willard Robinson from Geneva; Charles Fournier and Alexander Brailow from Gold Seal; Greyton Taylor from Taylor's; Adhemar de Chaunac and George Hostetter from Bright's; and others were very important in this movement. Herman Weimer and Herman Amberg were both very much involved with the New York industry. Both are still active in the eastern wine industry, but with strong emphasis on vinifera varieties.

The hybrids spread rapidly through the Finger Lakes between the 1940s and 1960s. The wineries in New York growing and making wine from the hybrids in the 1940s were Taylor, Widmer, Gold Seal, and Great Western. Taylor winery was the early leader but Widmer and

FIGURE 2.
Tasting at Taylor Wine Company, July 13, 1951. Standing (left to right): Waitress?, Jocelyn Wagner, Charles Fournier, Ludwig (attorney), X, Greyton Taylor, X, Ferguson. Sitting (left table, left to right): X, X, George Oberle, John Einset, Richard Wellington, Tony Doherty, Hector Carveth. Sitting (right table, left to right): E. Carleton, Willard Robinson, Adhemar de Chaunac, chemist from Chateau Gay, Alexander Brailow, Philip Wagner, Harry Hatch (owner of Brights Winery).
X = Unidentified people, agents from Washington, D.C. or friends of Greyton Taylor.

Gold Seal followed quickly. By 1963, Taylor Wine Company was making wine from more than 200 acres of 'Seibel 5279,' 43 acres in company-owned vineyards. One of the problems with the hybrids, Seaton (Zeke) Mendall of Taylor's recalls was bunch rot (19)—not black rot, but bunch rot, where the berries crack and just turn into a moldy, rotten mess. 'Seibel 7053' did not grow well in the Finger Lakes, but grew better in western New York. Grower comments were, "You could spray everyday and still get wiped out by the downy mildew."

'Seibel 9549' looked like a real winner to many wineries, growers, and researchers. The Canadians and Boordy Vineyards were the chief proponents for its culture. As far as can be determined, the source of vines for all the 'Seibel 9549' vineyards was from the Canadian source. The original 'Seibel 9549' vines were at the Vineland substation's vineyards until 1990. It became the third most-planted variety in Ontario, reaching 2,100 acres in 1981 (5, 14). Catell (9) states that 'Seibel 1000' was apparently the first hybrid introduced into Canada.

The first French hybrids introduced into Ohio were cuttings of 'Seibel 1000' by Mantey Winery in Sandusky in 1941 (18). They were then sent to Foster Nursery to be grafted onto Couderc 3309. Additional vines of 'Baco #1' were planted in 1955. The plantings consisted of 1.38 acres of 'Seibel 1000' and 1.3 acres of 'Baco #1.' Both plantings still exist at Firelands Winery (formerly Mantey Winery). Norman Mantey, one of three brothers and owner of Mon Ami Winery in Port Clinton, made a 'Pink Rose' wine from the 'Seibel 1000.'

In 1954, Meiers Wine Cellars obtained 'Baco #1,' 'Seibel 1096,' 'Seibel 5898,' and 'Seibel 4643' from Boordy Vineyards and planted them on North Bass Island (8). In 1962, they added 'Seibel 10878' and 'Seibel 7053,' also from Boordy. Additional 'Baco' vines were planted in 1968. Two rows of the original 'Baco' still exist. Myron Baker, Morrow, Ohio, planted 'Seibel 5279' in 1960. Vines were furnished by and fruit sold to Meier's Wine Cellars, Cincinnati, Ohio.

The Ohio Agricultural Research and Development Center planted 'Seibel 1000' in their vineyards at Wooster in 1956. In 1960, the first in a long series of hybrid selections were planted in research vineyards at the Southern Branch, Ripley, Ohio.

H.C. Barrett, a fruit breeder at University of Illinois, obtained a number of French hybrids in 1947-1949 through the USDA Plant Introduction Station in Maryland (2, 3). They were evaluated both for their value as new varieties as well as breeding parents.

Emmett Schroeder in Hutchinson, Kansas (25), who characterizes himself as an amateur grape grower, evaluated a number of French hybrids, not for wine but as table grapes. As far as can be determined, he obtained many of his selections from France between 1947 and 1949 through his church contacts (3).

Virginia Polytechnic Institute had introduced hybrids into their grape program as early as 1950, and by 1953, was reporting their potential use and adaptation for Virginia (20).

The first French hybrid vineyard planted in Pennsylvania was in 1953 by Melvin Gordon at

Birchrunville. Nursery stock was purchased from Boordy Vineyards (97).

Industry expansion

Availability and use of grapes from California during Prohibition and the disruption of wine from France during World War II both created an interest in and demand for non-foxy wine grapes in the eastern United States. Wineries couldn't supply the demand and started increasing production, which required serious vineyard expansion programs. In the Finger Lakes, wineries met some serious resistance from growers (19). They couldn't get the growers to put new vines in the ground. To solve the problem, Gold Seal, Taylor, Great Western, and Widmer had to buy land and establish their own plantings. Taylor's would plant a whole farm or several farms to 'Delaware,' 'Ives,' hybrids, or anything they could get. When Seaton Mendall went to work at Taylor's in 1943, they had seven acres of grapes; 10 years later, approximately 900 acres had been planted. But nobody was paying attention to soil conditions or other such details. It took several years to meet the demand.

Prior to the war, there were no serious grower relations programs among any of the wineries in New York. The growers suffered under the processors. But, when the war started and imports were cut off, the wineries decided they had to treat the growers better. Burt Taylor decided that quality control was a problem and they needed an "in-field quality control program" to improve their grower relations. Bad fruit was sent to the winery and had to be turned down.

In 1943, Mendall started a one-man operation trying to coordinate with growers and get them to do a better job (19). Mendall was in charge of vineyard operations for 10 years. As the "troubleshooter," he was in close contact with each grower two or three times a year. When harvest time came, he knew what each grower had in his vineyard and its quality. Thus, Taylor's eliminated the inspection problem at the plant. If something questionable did come in, Mendall just called on the phone to straighten it out. Still, only verbal contracts were given. Wineries paid on time and treated growers right, but written contracts were not given until years later. Tom Chadwick joined Mendall and took over when he retired from Taylor's in 1979.

Mendall's first experience with French hybrids was in 1943 when Taylor's got some hybrids from Philip Wagner at Boordy Vineyards. The first variety obtained was 'Seibel 5279.' Some 'Seibel 1000' vines were obtained about the same time, but they soon went out. Taylor's planted a large acreage of 'Seibel 5279' (200 acres by 1963) and used it for blending purposes.

"Unfortunately, it had to be planted on pretty good soil. When it was planted on some of the shallow clay soils with a foot to bedrock, the vines just did nothing. They almost stood still. If you got a pound per vine on them, you were lucky. You had to have a good soil" (19). Mendall still thinks that 'Seibel 5279' is a good blending wine.

In the late 1940s and early 1950s, everyone was very optimistic about wine. At Gold Seal, they were growing some hybrids and were

To compound the problem, the wine market changed from red to white! The growers were wild.

FIGURE 3.
Tasting at Pleasant Valley Wine Co. (Great Western), July 28, 1952. Standing (left to right): X, X, X, X, Hector Carveth (Chateau Gay Winery), X, George Oberle, E. Carleton, Chemist (Chateau Gay), John Einset, X, X, X, Alexander Brailow, Tony Doherty. Sitting (left to nght): Philip Wagner, Ludwig, Charles Fournier, X, Adhemar de Chaunac, Josylin Wagner, ureyton Taylor, X, X.
X = *Unidentified people, agents from Washington, D.C. or friends of Greyton Taylor.*

recommending varieties that they wanted the grape growers to plant. The wine business was changing rapidly. Growers were selling their American varieties but also got stuck with some when varietal demands changed. In 1953-54, the most popular wine was pink 'Catawba.' Actually it was barely over 50% real 'Catawba,' because quite a few hybrids were blended into it.

In general, the trend during this period was to try to make wines that appealed to the public, rather than adhere to an earlier philosophy: "This is what we make; you like it, you buy it!" At one time, this policy was successful, but it wasn't anymore, because wine consumers were again buying a significant quantity of imported wines.

Wineries had great hopes for 'Seibel 9549' from the standpoint of giving them a red blending wine which could possibly replace 'Ives.' Unfortunately, Tom Chadwick and Mendall at Taylor's didn't have any land to plant their 'Seibel 9549' on except old vineyard sites. Fred Taylor insisted they use the old sites. The vines were all planted on their own roots and on heavy clay soil. Vines grew well and as a general estimate produced about five tons per acre on six- to eight-year vines. They then decided to urge growers to plant them, as they needed more grapes every year. Red wines were much more predominant than white. So, when growers put in a vineyard, they told them to plant 'Seibel 9549.'

Unfortunately, the growers were putting these on new sites where a vineyard had not been before and on better soils. But when these vineyards came into production and were about five years old, they were producing 10 to 12 tons per acre, and Taylor's was anticipating a production of five to six tons. They were flooded with grapes. To compound the problem, the wine market changed from red to white! "The growers were wild."

O.A. Bradt at the Horticultural Institute at Vineland, Ontario, Canada, started cluster thinning research in 1949 when the hybrid introductions began to fruit. This work was continued for several years. The benefits were reported in appropriate publications and at many grower field meetings. Growers in New York didn't cluster-thin any of the hybrids until 'Seyve-Villard 5276' came along.

Gold Seal did not go too heavily into the planting of hybrids. Dr. Konstantin Frank colored the picture to a great extent for them. He was one of the greatest opponents to the use of hybrids in the East. Alexander Brailow, wine maker at Gold Seal since 1946, brought him from New York (7). Prior to this, Frank was working as a dishwasher in New York City. He barely knew English. He wrote to Professor Richard Wellington with whom Alexander Brailow had worked at the New York Agricultural Experiment Station.

Professor Wellington knew that Brailow knew Russian and put him in touch with him. Brailow invited Frank to come to Geneva. Nelson Shaulis obtained work for him at the Geneva Experiment Station working with blueberries, but he wasn't satisfied there. Eventually, Brailow helped him get a job at Gold Seal as vineyard manager. Charles Fournier was president of Gold Seal at this time. Frank was very opinionated and difficult to work with, but was a good vineyard manager. "No hybrids, no labruscas, pull everything out and plant viniferas, now!" Frank preached. According to numerous sources (7, 19), Frank created problems for the whole French hybrid program at Gold Seal. He hated the French hybrids. He had no use for them whatsoever. He eventually built and started his own winery in 1962, in Hammondsport, New York, and attained quite a reputation. Frank died in 1985. His son, Willie, now owns the winery and is again making wine.

Concerning the hybrids, Frank expounded that they had a smooth leaf which was attractive to the aerial form of phylloxera. The proliferation of the aerial form completed the normal cycle and thus increased the intensity of the phylloxera population. By contrast, the American varieties were not attractive to aerial phylloxera and, therefore, it led a limited (halfhearted) existence only on the roots, thus enabling these vineyards to survive without severe phylloxera damage. This philosophy was expounded in an unpublished paper entitled "The Present Vista for the Vitis Vinifera European Grape Varieties in the East" (15). According to Frank, all French hybrids were destroyed in Gold Seal vineyards by 1957.

A controversy also raged between Frank and the New York wine industry that the French hybrids contained wood alcohol (methyl hydrate). Also, there were other "biostatic" (toxic) substances in French and American hybrids which, when fed to chickens, produced crippled offspring. This information was based on experiments conducted by Hans Breider in Wurtzburg, Germany (27). Gil Stoewsand was hired by the New York Agricultural Experiment Station,

> *"No hybrids, no labruscas, pull everything out and plant viniferas, now!"*
>
> Dr. Konstantin Frank

Geneva, to determine if there were such problems with hybrid wines. His investigations proved conclusively that it was the chicken ration itself that was deficient in vitamins and other components, and not the wine made from the hybrids.

The overall quality of wines in the East was improved during this period by the use of the hybrids and provided the wineries with some different wine types. Taylor's Lake Country series became the leader in the business and sold extremely well. Varietal wines gradually became popular with Taylor's after they acquired Great Western and the Pleasant Valley Wine Company in 1962.

Taylor's was first to establish prices on grapes. This eventually resulted in their setting the grape prices for the whole region. That policy remained for a number of years. About 1975, a law was passed in the state that processors had to notify their growers on or before August 15 what prices were going to be. Everybody eventually settled for an August date.

French hybrids named

Prior to 1970, names could be found for several of the hybrids and were freely used interchangeably with the numbers. But growers and wineries were not very comfortable with this procedure. How official some of these names were or where they came from was not broadly understood. Some of them had received names in France, both from popular use and by official organizations. In the United States, names could be found in several authoritative publications. For example, the Register of New Fruit and Nut Varieties, published periodically in the *Proceedings of the American Society for Horticultural Science* and later in *HortScience*, listed synonyms for six hybrid cultivars—three from Seibel and three from Seyve-Villard. They are as follows:

'Seibel 5279': 'Aurore' (Vol. 74, 1959)
'Seyve-Villard 20-473': 'Muscat de Saint Vallier' (Vol. 76, 1960)
'Seibel 4643': 'LeSequanois' and 'Roi des Noirs' (Vol. 81, 1962)
'Seibel 4986': 'Rayon d'Or' and 'Roi des Blances' (Vol. 81, 1962)
'Seyve-Villard 20-365': 'Dattier de St. Vallier' (Vol. 87, 1965)
Seyve-Villard 20-366': 'Pierrelle' (Vol. 87, 1965)

Two New York publications in 1965 (24) and 1968 (1) list names for 11 additional selections. They are:

'Kuhlmann 188-2': 'Marechal Foch'
'Seibel 5455': 'Plantet'
'Seyve-Villard 23-410': 'Valerien B'
'Baco #1': 'Baco Noir'
'Kuhlmann 188-2': 'Foch'
'Landot 244': 'Landal'
'Seibel 4986': 'Rayon d'Or'
'Seibel 5279': 'Aurora'
'Seibel 8357': 'Colobel'
'Seibel 10878': 'Chelois'
'Seyve-Villard 12375': 'Villard Blanc'

The naming of an additional seven previously unnamed hybrids was proposed in 1971 and eventually accomplished under the authorization of the Great Lakes Nomenclature Committee in

1973, a committee representing five groups in the northeast United States and Canada, organized at Vineland, Ontario, in January 19, 1972 (5).

Early in 1970, Seaton Mendall of Taylor's made a call to Harold Olmo at the University of California at Davis (19). Olmo was then registrar of New Fruit and Nut Varieties. Mendall told him that the Finger Lakes Wine Growers Association (FLWGA) wanted to name some of the previously unnamed French hybrids. Further, they were going to get together with the processors and ask each one of them to submit names that they would like to have these different numbered hybrid selections called. This action caused considerable concern both in and outside of the state of New York.

Mendall, on behalf of the FLWGA then contacted the American Pomological Society and told them that there was a group of French hybrid wine varieties becoming important in New York as well as other states, and wanted to name them. They thought seven selections had not been named by the French originators, and they felt would never be sufficiently important to any of the French districts to be named. They were further told that New York and other eastern and midwestern wineries were going to make varietal wines out of these French hybrid grapes and felt it would be difficult to sell a bottle of wine with number (S.9549, S.10878, etc.) on the label. Therefore, the FLWGA was going to name them. Mendall attended the American Society of Horticulture Science/American Pomological Society annual meeting in Miami, Florida, in November 1970 and presented their proposal at a general session. The American Pomological Society Nomenclature Committee gave approval to go ahead but to select some short names.

It all came to a head at a meeting at Taylor's winery with the processors. As a result of discussing, arguing, and voting, a decision was finally reached. The names of seven selections were published by the association in an eight-page leaflet dated August 24, 1970, entitled "French-American Hybrid Names" (22). Descriptions were supplied by George Remaily and George Slate of the New York Agricultural Experiment Station, Geneva, New York. However, both Olmo and the Canadians were concerned that recognized procedures were not being followed. In a letter to N.H. Paul, president of Finger Lakes Wine Growers Association, from E.A. Kerr, chairman of an ad hoc committee of the Research Section of the Research and Development Committee of the Canadian Wine Institute, on July 23, 1971, indicated that international rules were being circumvented, since FLWGA was not conforming to the International Code of Nomenclature of Cultivated Plants of 1969.

In a letter to Gerard Marot in March 1970, George Slate listed six (not seven) hybrids that had not been named ('Seibel 1000,' '5898,' '7053,' '9549,' '9110,' and '13053'). Marot replied that they were named and he should write to R. Seibel (Albert Seibel's nephew). A letter from Seibel, dated April 1970, suggests that only four of the seven were named, and he preferred the acceptance of these, but names for the others could be made by the committee (FLWGA or the Great Lakes

It all came to a head at a meeting at Taylor's winery with the processors.

Grape Nomenclature Committee) and should be accepted in North America and published in the Register of New Fruit and Nut Varieties. The names of the four were:

'Seibel 7053': 'Perle d' Aubenas'
'Seibel 9110': 'Clairette Seibel'
'Seibel 13053': 'Noir precoce de Seibel'
'Seibel 5898': 'Seibel Noir'

Much correspondence and many meetings by interested parties took place between 1970 and 1972. As a result of a meeting at Rochester, New York, on January 19, 1972, a committee was set up, called the Great Lakes Grape Nomenclature Committee. Five organizations were to be represented on the committee. In March 1972, the committee was organized with O.A. Bradt from the Horticultural Research Institute of Ontario, Vineland, as president and representative of the Great Lakes Grape Research Coordinating Committee. Other members of the committee were: Ernest A. Kerr, Horticultural Research Institute of Ontario, Vineland; John Einset, representing the New York Agricultural Experiment Station; and Seaton Mendall from the FLWGA (5).

As a result of the January meeting a letter was written by O.A. Bradt to H.P. Olmo, registrar of New Fruit and Nut Varieties, on March 30, 1972, requesting that, with one exception, the seven varieties, previously published by the Finger Lakes Wine Growers Association, be approved and be officially accepted and recorded. The exception was 'Cameo' ('Seibel 9549'), which the Canadians said did not fit this grape. 'Cameo' meant something else, a lighter red wine; a light (white) yellow wine. Adhemar de Chaunac had done more than anybody else with it in its introduction and winemaking. In the end, the grape was named 'De Chaunac' in honor of Adhemar de Chaunac. The official names were: 'Rosette' ('Seibel 1000'), 'Rougeon' ('Seibel 5898'), 'Chancellor' ('Seibel 7053'), 'Verdelet' ('Seibel 19110'), 'De Chaunac' ('Seibel 9549'), 'Cascade' ('Seibel 13053'), 'Vignoles' ('Ravat 51').

A copy of this letter to H.P. Olmo was also sent to Gerard Marot, president of the Federation National de la Viticultural Nouvelle in France.

In 1973, an article was published by O.A. Bradt and J. Einset in the *Fruit Varieties Journal*, with names of the seven French hybrids (6). As for the concerns of the French breeders, there was some correspondence, and finally Seibel's nephew was told that the varieties had been officially named in the U.S. No additional correspondence is available to indicate further interchange of information on the subject. In all probability, it was Charles Fournier, President of Gold Seal Winery, who acted as the intermediary, since all correspondence from them was in French.

French hybrids today

Between the 1940s and the 1980s, hundreds of hybrids were evaluated by growers, wineries, and research institutions in states mostly east of the Rocky Mountains, Ontario, and British Columbia, Canada. From this base has come the production of hundreds of acres of French hybrids (HPDs) and a significant volume of quality wine. The hybrids are still used for blending, as well as for a respectable number of varietal wines. Most significant has been the large number of wineries, many small by comparison to California, that have come into existence because of the popularity of the hybrids. The freedom to depart from traditional eastern wines and produce new, more neutral, and broadly accepted table wines may have been the key to the transition that has occurred.

The most prominent varieties to emerge through this 40-year developmental period are (not in order of total acreage or importance):

White: 'Seyval Blanc' ('S.V.5276'), 'Vidal Blanc' ('Vidal 256'), 'Vignoles', ('Ravat 51'), 'Aurore' ('S.5279'), 'Villard Blanc' ('S.V. 12-275').

Red: 'Baco Noir' ('Baco #1'), 'Chambourcin' ('Joannes Seyve 26-205'), 'Marechal Foch' ('Kuhlmann 188-2') , 'Chancellor' ('S.7053'), 'De Chauna,' ('S.9549'), 'Chelois' ('S.10878'), 'Leon Millot' ('Kuhlmann 194-2'), 'Rosette' ('S.1000'), 'Rougeon' ('S.5895'), and 'Villard Noir' ('S.V.18-315').

By 1981, the acreage of 'De Chaunac' in the province of Ontario, Canada, made it the third most prolific variety, with a peak production of 2,100 acres. However, beginning in 1988, a major transition began in the grape and wine industry in Ontario, Canada, and brought about many changes in variety preferences. Acreage of 'De Chaunac' has been greatly decreased, while 'Vidal' has emerged as an outstanding variety for ice wine and table wine. Varieties of grapes now authorized by the Vintners Quality Alliance for appellation of origin in Ontario include 11 French hybrids. Another seven French hybrids may be used in blends (11, 15). There has also been a major shift to *V. vinifera* acreage. Since 1986, the number of vines planted to vinifera has grown from 955,000 to 3.2 million in 1995.

Varietals for wine: 'Baco Noir,' 'Castel,' 'Chambourcin,' 'Chancellor,' 'Chelois,' 'Muscat du Moulin,' 'Leon Millot,' 'Marechal Foch,' 'Seyval Blanc,' 'Vidal Blanc,' and 'Villar Noir.'

Varietals for blending only: 'Aurore,' 'De Chaunac,' 'Landal,' 'Verdelet,' 'Vignoles,' 'S.V.23.572,' and 'J.S.23.416.'

In the eastern United States, there are similar but not so dramatic shifts toward the planting of vinifera varieties. Although the production of grapes from vinifera is still guaranteed to provide severe challenges from climatic extremes, the demand from wineries for them is providing good economic opportunities. One thing for certain, the connotations of hybridization as exemplified by the words 'French hybrid' will be with us for many years to come. Breeding programs in Germany and North America have incorporated them into their programs. Examination of new varieties created by current breeding programs will frequently identify one or more of them in the parentage. The French, to avoid the negative association of the term 'hybrid,' use the word 'metis,' which connotes the crossing of pure *V. vinifera* varieties (21).

The hybrids are still used for blending, as well as for a respectable number of varietal wines.

Literature cited

1) Anon. 1968. A Catalog of New and Noteworthy Fruits. N. Y. St. Frt. Test. Coop. Assoc. Geneva, NY.

2) Barrett, H.C. 1956. The French Hybrids. *National Horticulture Magazine.* 35:132-144.

3) 1996. Personal communications.

4) Bradt, O.A. 1948. The Grape in Ontario. Hort. Expt. Sta., Ontario Dept. of Agr. Bul. 487, Toronto. pp. 31-32.

5) 1996. Personal communications.

6) Bradt, O.A. and John Einset. 1973. French hybrid grapes given names. *Fruit Var. Jour.* 27:58-59.

7) Brailow, Alexander. 1991. Personal communications.

8) Burris, Dale. 1996. Personal communications.

9) Cattell, H. and H.L. Stauffer. 1978. Wines of the East: The Hybrids. *Wine East.* Lancaster, PA.

10) Cattell, H. 1986. The pioneering years at Brights. *Wine East.* Lancaster, PA. 14:10-11, 26-27.

11) Cattell, H. 1995. A new quality image for Ontario wines. *Wine East.* Lancaster, PA. 23:4-10, 39.

12) Dunstan, R.T. 1962. Vinifera type grapes for the east. *Fruit Var. Jour.* 17:6-8.

13) Einset, J. and C. Pratt. 1975. The French Hybrid Grapes. In: *Advances in Fruit Breeding.* Janick, J., and J. Moore (eds.). Purdue Univ. Press.

14) Fisher, H. 1996. Personal communications.

15) Frank, Konstantin. The present vista for the *Vitis vinifera* European grape varieties in the east (in-house mimeo).

16) Galet, Pierre. 1979. A Practical Ampelography, Translated by Lucie T. Morton. Commstock Publishing Associates, Cornell Univ. Press. Ithaca, NY. pp. 154-186.

17) 1988. Les Porte-Greffes: Chapter IV, p. 209. Les Hybrides Producteurs Directs: Vol. V, pp. 369-380. In: *Cépages et vignobles de France.* 2nd ed. Publ. C. Dehan, Montpellier.

18) Mantey, Don. 1996. Personal communications.

19) Mendall, Seaton. July 13, 1991. Personal communications.

20) Moore, R.C. and G.D. Oberle. 1953. French hybrid grapes in Virginia. *Fruit. Var. Jour.* 8:5-8.

21) Olmo, H. 1961. (Abstracts and Reviews) Amer. Soc. Enol. & Vit. 12:192-93. *From:* Fénavino, Bull. officiel de la Fédération Nationale de la Viticulure Nouvelle et du synducat des hybrideurs et métissurs viticoles 14: No. 155, 34-56 (1961).

22) Paul, N.H. 1970. French-American Hybrid Names. The Finger Lakes Wine Growers Association. Naples, NY. Aug. 24, 1970. pp. 1-6.

23) Riley, C.V. 1883. The grape Phylloxera. *In:* Bushberg Catalog. Publ. by Bush & Son & Meissner. *Grape Growers Manual.* 3rd edition. pp. 52-56.

24) Robinson, W.B., J. Einset, K.H. Kimball, and J.J. Bertino. 1966. Notes on 1965 wine samples. Res. Circ. No. 7. New York Agricultural Experiment Station, Geneva, NY.

25) Schroeder, Emmett. 1950. French hybrids as table grapes. *Fruit. Var. Jour.* 5:85-90.

26) Snyder, Elmer. 1937. Grape development and improvement. In: *United States Department of Agriculture Yearbook.* pp. 631-664.

27) Stoewsand, G. 1996. Personal communications.

28) Wagner, Philip M. 1992. Personal communications.

29) 1978. Honorary lecture American Society of Enologists, Twenty-Ninth Annual Meeting, San Diego, California.

30) 1955. The French Hybrids. *Amer. Soc. Enol. & Vit.* 6:10-17.

Author Cahoon was with the Department of Horticulture and Crop Science, Ohio Agricultural Research and Development Center, Ohio State University, when this article originally appeared, October 1996.

The 'Desirable' Pecan

by Darrell Sparks

There are four cultivars that have become the standards of the pecan industry. They are 'Desirable,' 'Schley,' 'Stuart,' and 'Western Schley.' These cultivars are planted in large acreages throughout the pecan regions of the world. 'Schley' and 'Stuart' were two of the original "big four" planted in the southeastern United States in the 1920s; the other two were 'Alley' and 'Pabst' (10). 'Desirable,' 'Schley,' 'Stuart,' and 'Western Schley' are all being used in newly planted orchards, although new plantings of 'Schley' are very limited. These cultivars became standards of the industry because they proved to be profitable over a wide range of conditions and remained profitable as mature trees. These cultivars have passed the severe test of time.

'Desirable' is supposedly one of the first pecan cultivars developed from a controlled cross. The cross was made in the early 1900s by Carl F. Forkert from Ocean Springs, Jackson County, Mississippi (10). The parentage is unknown (6), but may be 'Success' x 'Jewett.' 'Desirable' was introduced about 1915 (18), but was not widely disseminated prior to Forkert's death in 1928. This cultivar would probably have been lost if scions had not been brought to the U.S. Pecan Field Station, Philema, Georgia, in 1925 (Fig.1). From this station, 'Desirable' was extensively disseminated as US-7191 for test planting beginning in 1930, was introduced commercially in 1945 (3), and was widely planted by the

A History of Fruit Varieties

FIGURE 1.
The 'Desirable' tree, Philema, Georgia. The tree, which is a top-worked 'Schley,' is one of the original 'Desirable' trees in Georgia. The tree is on the grounds of what was once the U.S. Pecan Field Station.

FIGURE 2.
The 'Desirable' leaf. The leaflets are reflexed relative to the horizontal plane of the rachis.

early 1960s. Mr. R.M. Marbury, Sr., was one of the first, if not the first, person to establish a 'Desirable' tree from the Philema Station source.

In the early 1930s, he topworked a tree in his yard on Gillionville Road, Albany, Georgia. Later, Marbury top-worked a portion of his orchard to 'Desirable' (F.G. Marbury, Sr., personal communication). This orchard, now known as Blue Three, is located a few miles south of Albany on Georgia Highway No. 19.

In the late 1960s and in the 1970s, planting of 'Desirable' decreased with the introduction of and renewed interest in the United States Department of Agriculture (USDA) cultivars. Presently, 'Desirable' plantings are on the increase which is associated with the less than anticipated performance of most of the USDA cultivars. Currently, in Georgia, 'Desirable' is the number one cultivar planted in new orchards. 'Desirable' has been used in pecan breeding, and two cultivars, 'Houma' and, probably, 'Kiowa,' have been released with 'Desirable' parentage.

In the southeastern United States, bud break in 'Desirable' is early and about five days before 'Stuart' (21), but the differential in bud break can be much greater in seasons following mild winters. The primary bud of 'Desirable' is unusually plump, roundish, and prominent. The roundish bud often abruptly terminates in a sharply pointed tip. These characteristics become more evident from the base to the tip of the shoot. The leaf is large with reflexed leaflets (Fig. 2). In most situations, the foliage tends to be light green with the pale color often, although not always, being a distinguishing characteristic. If both leaf nitrogen and potassium are at sufficient levels, the foliage is dark green (21). Leaf retention in the fall is good (11), but color retention is poor during prolonged cool autumns. As a young tree, shoot growth is vigorous. Development of shoots is restricted to the apical portion of the one-year-old branch. That portion of the one-year-old branch without shoots is sometimes called "blind" wood. Restriction of shoots to the apical portion of the one-year-old branch results in a fan as opposed to a pole branching habit (21). The tree is moderately open (Fig. 3). Tree form is more spreading and open than 'Stuart' (1). The open nature of the tree reduces shading out in the interior of the canopy, which greatly increases the fruiting area of the tree. The tree grows vigorously and, consequently, is not suited to high density planting. 'Desirable' is moderately difficult to propagate.

'Desirable' is protandrous. An abundance of pollen is produced very early in the pollinating season, and 'Desirable' is an excellent early pollinizer for many cultivars over a wide geographic range. Occasionally, the prolific catkin production is associated with death of a few one-year-old branches within a tree. Branch death results from excessive depletion of carbohydrate reserves by the massive catkin production. Extensive overlapping of pollen shedding and stigma receptivity is common in areas with mild winters (21). In Brownwood, Texas (13), and Brazil (7), 'Elliott' is a suitable pollinizer for 'Desirable;' in Georgia (27), 'Cape Fear' and 'Elliott' in combination will pollinate 'Desirable.'

'Desirable' is more precocious than 'Stuart' and comes into commercial production about two years earlier. Under ideal culture, a commercial yield is obtained in the sixth year. The yield from a 'Desirable' tree is about 15% greater than 'Stuart' (21). 'Desirable,' like 'Stuart,' has a "built-in" fruit-thinning mechanism. Thinning occurs during the second drop and is due to a lack of pollination and/or fertilization, either one or both of which do not occur in many of the flowers. Lack of fertilization is, by far, more likely (22). Regardless, a second fruit drop occurs consistently from one year to the next. The number of fruits per cluster is often only one or two. The inherent capacity for fruit thinning and the open nature of the tree are dominant factors in 'Desirable' being one of the, if not the, most consistent cultivars in annual production. Occasionally, however, the second drop is excessive and yields are substantially below normal. On the other extreme, when the second drop is low, the tree fruits excessively and production is off the next year.

The shuck of the fruit is glossy and convoluted (Fig. 4). Nut maturity is about four days later than 'Stuart' (21). However, nut maturity in 'Desirable,' unlike 'Stuart,' is not excessively staggered, and a higher percentage of fruit can be removed from the tree during the first harvest.

Nut shape is oblong or blocky with an obtuse base and an apex that varies from obtuse to obtuse asymmetric. When the apex is asymmetric, the apex has the appearance of being slightly curved. Shell halves within nuts vary from equal to unequal in size. Cross section is oval, and the nut is slightly compressed on the nonsuture sides. The surface of the compressed areas is rough, otherwise shell topography is smooth. The suture is sometimes slightly elevated, but not throughout its entire length. Ridges are subtle, but often there is one ridge which is more evident near the apex. Stripes are sparse. Dots are more abundant than stripes, but are not dense. Markings are brownish-black on a light brown background (21).

The 'Desirable' nut is large and bigger than 'Stuart.' A high quality 'Desirable' will have a count of 42 to 44 nuts per pound, although the average is more like 47. The shell is medium thick and thinner than 'Stuart.' As a result of the medium thick shell, there is minimal breakage during mechanical harvesting. The nut is also very suitable to mechanical shelling, and a high percentage of intact halves is obtained. A percent kernel of 52 is good, with 54 being excellent. If percentage kernel is on the low side, for example 52%, the interior of the kernel is solid. This is in contrast to 'Stuart,' which tends to have "air pockets" when percentage kernel is marginally low. The kernel is very attractive with a color rating of 6.8 (1 = dark; 10 = light) or higher. Flavor is good and is considered by some to be better than 'Stuart' (21).

Like 'Stuart,' 'Desirable' tends to produce good quality nuts during a heavy crop year. This is due to the fact that the number of nuts in a cluster is not excessively large. There is a strong demand for 'Desirable' in the marketplace. The demand is due to the relatively large kernel size, good kernel color, consistent quality, and ease of hand cracking (when sold in-shell). 'Desirable'

FIGURE 3.

Tree form in 'Desirable.' The tree is moderately open due to the relatively sparse branching and fairly wide crotch angles.

FIGURE 4.

'Desirable' fruit cluster size often consists of one or two fruits. Shuck surface is glossy and convoluted.

FIGURE 5.

Classical symptom of nitrogen and potassium imbalance in 'Desirable.' The scorch usually first appears at the base of the leaflet, as indicated in the photograph. The basal two leaflets are lighter in color than the next two leaflets. The lighter color is indicative of potassium deficiency. In Georgia, the scorch occurs in mid-May if the disorder is very severe, but normally it occurs in early to mid-June. Defoliation can be massive. Scorch from a nitrogen and potassium imbalance occurs before fungal leaf scorch (21). Also, the scorch is not bordered by a dark line, as with fungal leaf scorch.

kernels are one of the best for roasting and salting. This is because the kernel has good color retention during the roasting process. 'Desirable' nuts typically sell for a higher price than 'Stuart.' In storage, kernel stability is good in-shell, but only fair if shelled (26).

'Desirable' breaks buds early in the spring, making the tree very susceptible to late spring freezes. 'Desirable' is also susceptible to early fall freezes (9) and to winter freezes (23, 24). Winter injury is especially serious in young trees. Consequently, 'Desirable' should not be planted in marginal climates or on poor air drainage sites.

As a general rule, seedling juvenile trunks, which are far more resistant to winter injury than cultivar trunks (23), should be used whenever 'Desirable' is planted in areas subject to freezing temperatures. The bud or graft union should be at least one foot above the ground, the greater the distance the better. Because weak trees are especially susceptible to winter injury, every attempt should be made to keep the tree healthy. Consequently, retaining the foliage on the tree until the end of the growing season is of major importance. Special attention should be given to the potassium status of the tree, as low potassium increases susceptibility of pecan trees to winter injury (17).

Abnormal flowering (male and female inflorescence within the pistillate cluster) occasionally occurs in 'Desirable.' Abnormal flowering is induced by temperatures near freezing during or just before bud break (20). Abnormal flowering occurs more often in 'Desirable' than most other cultivars and may be due to the earliness of bud break, increasing the chances of winter injury rather than any inherent characteristic. Abnormal flowering does not occur frequently and thus is not a major detriment to production.

'Desirable's' susceptibility to scab varies with geographical location. At Hanna, Louisiana, 'Desirable' is much more susceptible to scab than 'Stuart' (1), but at Gainesville, Florida, the reverse occurs (16). In Louisiana, 'Desirable' was reported to scab as severely as 'Schley' (1) and 'Western Schley' (15) in the 1970s, but it was immune in 1944 (8). In areas where 'Desirable' is highly susceptible to scab, experience has shown that scab is much easier to control in the open canopy 'Desirable' than in the dense canopy 'Schley.' The leaf is also susceptible to vein spot, downy spot, and liver spot (8). The fruit is especially susceptible to powdery mildew (2). Control, however, of all these diseases is relatively easy if sprays are timed properly. This cultivar has moderate resistance to fungal leaf scorch (12) but high susceptibility to bunch disease (11).

'Desirable' is immune to southern pecan leaf phylloxera at Brownwood, Texas (4), but is susceptible to pecan phylloxera (1). The cultivar has moderate resistance to pecan bud moth (14), black pecan aphids (5, 21), and stink bug (5). Following yellow aphid infestations, the leaf has good resistance to sooty mold accumulation (21, 25). 'Desirable' is very susceptible to the potato leafhopper. Severe leaf curl results from heavy infestations occurring during the leaf expansion period (April and early May in the southeastern United States) or during the period of the second cycle of shoot growth. In addition to damage to the leaf, this insect poses a problem of early insecticide use versus integrated pest management. The necessity to spray for potato leafhopper may kill beneficial insects which repress other insect pests, mainly aphids.

'Desirable' is very sensitive to an imbalance between nitrogen and potassium (Fig. 5). Potassium must be maintained at a relatively high level in the leaf, or leaf scorch and premature defoliation occur in mid-May or June (Georgia conditions). In a planting of mixed cultivars, the scorch is an excellent indicator of the overall potassium status of the orchard. If scorch does not occur in 'Desirable,' potassium is within the range for high yield and nut quality in 'Desirable' as well as in other cultivars in the orchard (19). Foliage retention is good (11) when nitrogen and potassium are properly balanced. The shoot is moderately susceptible to mouse ear.

'Desirable' has an especially high sunlight requirement. The high light required by 'Desirable' is particularly evident when this cultivar is used as a replant in a mature orchard. Tree growth is often weak and spindly, relative to a 'Stuart' replant. The growth problem is further accentuated by 'Desirable's' sensitivity to a nitrogen-potassium imbalance, which requires extra potassium application above that given to mature trees. 'Stuart' does not have these problems to the same degree as 'Desirable,' and, consequently, 'Stuart' makes a better replant choice in an old orchard.

Mature 'Desirable' trees are about two and one-half times more sensitive to high winds than 'Stuart' (21). Consequently, 'Desirable' definitely should not be planted in coastal areas that are subject to hurricanes. 'Desirable's' sensitivity to wind damage is due, in part, to the tree's strong tendency to produce weak limb crotches. During the tree's early life, training is essential for a strong tree structure because of the tendency to produce weak crotches.

'Desirable' is sensitive to drought conditions. During water stress, leaves are lost more readily from 'Desirable' than from some other cultivars such as 'Elliott,' 'Farley,' and 'Stuart,' On the other extreme, 'Desirable' is more sensitive than most other cultivars to poorly drained soils.

'Desirable' is an excellent cultivar because of the production of better than average yields on a fairly consistent basis, good nut size, dependable kernel quality, high suitability to mechanical shelling, and high demand for the nut in the marketplace. As a mature tree, 'Desirable' is probably the best cultivar available. However, 'Desirable' is one of the most sensitive cultivars and should not be planted unless excellent cultural practices are to be employed.

> *As a mature tree, 'Desirable' is probably the best cultivar available.*

Literature cited

1) Anon. 1970. Dooryard pecan trees in Louisiana. Circular letter to homeowner. USDA Pecan Lab., Shreveport, Louisiana.

2) Bertrand, P.E. 1984. Powdery mildew. *Proc. Georgia Pecan Growers Assn.* 19:7-9.

3) Brooks, R.M. and H.P. Olmo. 1951. Register of new fruit and nut varieties: List 6. *Proc. Amer. Soc. Hort. Sci.* 58:386-407.

4) Calcote, V.R. 1983. Southern pecan leaf phylloxera (Homoptera: Phylloxeridae): Clonal resistance and technique for evaluation. *Environ. Entomol.* 12:916-918.

5) Calcote, V.R. and W.A. Scott. 1988. Pecan variety performance in the Mississippi Delta. *Proc. Southeastern Pecan Growers Assn.* 81:83-95.

6) Crane, H.L., C.A. Reed, and M.N. Wood. 1938. Nut breeding, pp. 827-887. *In:* G. Hambige (ed.). *Yearbook of Agriculture 1937,* Gov. Print. Off., Washington, D.C.

7) Da Costa Baracuhy, J.B. 1980. Determinagao do periodo de fioragao e viabilidade do pollen de differentes cultivares de noqueira pega, *Carya illinoenis* (Wang.) K. Koch. Ph.D. Diss. Universidade Federal De Pelotas, Pelotas, Brazil.

8) Dodge, E.N. 1944. Pecan varieties. *Proc. Texas Pecan Growers Assn.* 23:7-16.

9) Goff, W.D. and T.W. Tyson. 1991. Fall freeze damage to 30 genotypes of young pecan trees. *Fruit Var. J.* 45:176-179.

10) KenKnight. G.E. 1970. Pecan varieties 'happen' in Jackson County, Mississippi. *Pecan Quarterly.* 4(3):6-7.

11) KenKnight, G. and I.H. Crow. 1967. Observations on susceptibility of pecan varieties to certain diseases at the Crow farm in DeSoto Parish, Louisiana. *Proc. Southeastern Pecan Growers Assn.* 60:48-57.

12) Littrell, R.H. and R.E. Worley. 1975. Relative susceptibility of pecan cultivars to fungal scorch and relationship to mineral composition of foliage. *Phytopathology* 65(6):717-718.

13) Madden, G.D. and E.J. Brown. 1975. Budbreak, blossom dates, nut maturity and length of growing season of the major varieties grown in the West. *Pecan South* 2(3):96 97, 112, 113.

14) Mizel, R.E. and D.E. Schifflhauer. 1986. Larval infestations levels of pecan bud moth *Gretchena boilinana* (Lepidoptera: Tortricidae), in relation to cultivar and position on the tree. *Environ. Entomol.* 15:436-438.

15) Sanderlin, R.S. 1987. Evaluation of pecan cultivars and USDA selections for scab disease susceptibility. Res. Rpt. Pecan Res-Ext. Sta. Louisiana Agr. Exp. Sta., Shreveport.

16) Sherman, W.B. and N. Gammon. 1977. Performance of pecan cultivars in north central Florida. *Pecan South* 4(1):38-40.

17) Sharpe, R.H., G.H. Blackmon, and N. Gammon, Jr. 1952. Relation of potash and phosphate fertilization to cold injury of Moore pecans. *Proc. Southeastern Pecan Growers Assn.* 45:81-85.

18) Smith, C.L. and L.D. Romberg. 1948. Pecan varieties and pecan breeding. *Proc. Texas Pecan Growers Assn.* 27:18-26.

19) Sparks, D. 1976. Nitrogen scorch and the pecan. *Pecan South* 3:500-501.

20) Sparks, D. 1992. Abnormal flowering in pecan associated with freezing temperature. *HortScience.* 27:801-803.

21) Sparks, D. 1992. Pecan cultivars—the orchard's foundation. Pecan Production Innovations. Watkinsville, Georgia.

22) Sparks, D. and G.D. Madden. 1985. Pistillate flower and fruit abortion in pecan as a function of cultivar, time, and pollination. *J. Amer. Soc. Hort. Sci.* 110:219-223.

23) Sparks, D. and J.A. Payne. 1977. Freeze injury susceptibility of nonjuvenile trunks in pecan. *HortScience* 12:497-498.

24) Sparks, D. and J.A. Payne. 1978. Winter injury in pecans—A review. *Pecan South.* 5(2):56-60, 82-88.

25. Sparks, D. and I. Yates. 1990. Pecan cultivar susceptibility to sooty mold related to leaf surface morphology. *J. Amer. Soc. Hort. Sci.* 116(1):6-9.

26. Woodroof, J.G. and E. K. Heaton. 1961. Pecans for processing. Georgia Agr. Exp. Bull. 80.

27. Worley, R.E., O.J. Woodard, and B. Mullinix. 1983. Pecan cultivar performance at the Coastal Plain Experiment Station. Univ. Georgia Agr. Exp. Sta. Res. Bull. 295.

Author Sparks was with the Department of Horticulture, University of Georgia, Athens, Georgia, when this article originally appeared, January 1997.

The 'Cortland' Apple

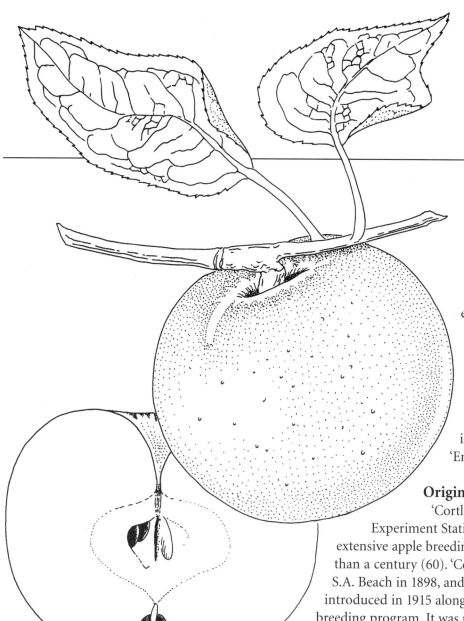

by Roger D. Way & Susan K. Brown

'Cortland' is one of the few apple cultivars of hundreds originating from controlled apple breeding programs early in the twentieth century that was good enough to become extensively grown on a commercial scale. In 1923, it was awarded the silver Wilder Medal by the American Pomological Society (58). By 1965, 50 years after its introduction, 'Cortland' had become the third most important cultivar grown in New York State, which was then the second most important apple state in the United States; 'McIntosh' and 'Rhode Island Greening' the two leading cultivars. Although its importance declined in the latter half of the century, in midcentury it was the most important apple cultivar introduced from Geneva; later, 'Jonagold' and 'Empire' surpassed it.

Origin

'Cortland' was bred and introduced at the New York State Agricultural Experiment Station, Cornell University, Geneva, New York. It was released from the extensive apple breeding program that began there in 1895 and has now continued for more than a century (60). 'Cortland' resulted from the cross 'Ben Davis' x 'McIntosh,' made by S.A. Beach in 1898, and was selected in 1911 from a population of 11 seedlings. 'Cortland' was introduced in 1915 along with five other apple cultivars. It was the first product of this apple breeding program. It was named for Cortland County, New York.

The breeding work that led to the origin of 'Cortland' was described by Hedrick and Wellington

FIGURE 1.
The original 'Cortland' apple tree, circa 1922. Photo by H.B. Tukey.

(28), and the introductory descriptions of 'Cortland' were first published by Hedrick (24, 25, 27). Nursery trees were first distributed to fruit growers in 1915 (24). In 1952, 'Cortland' was briefly described in the Register of New Fruit and Nut Varieties (12).

Tree

Shaw (52) described in minute detail the vegetative characteristics of younger 'Cortland' nursery trees so they could be distinguished from other cultivars.

Mature 'Cortland' trees are large, vigorous, spreading, drooping, medium dense, round-topped, annual bearing, and very productive. The trunk is stocky, brownish gray, and uncommonly smooth, with very little shaggy bark. Because of its terminal bearing habit, many forked shoots occur, and the twigs are long and thin, almost willowy. As a result of this willowy growth habit, 'Cortland' trees are more difficult to prune than trees of most other apple cultivars.

Winter hardiness and frost resistance

'Cortland' trees are resistant to severe winter cold. In New York, following the severe winter of 1933-34, trees of 'Cortland,' like its parent 'McIntosh', survived without injury (1, 2). At Ottawa in the same winter, 25% of the branches on 'Cortland' trees were killed by the severe cold (9). In eastern Canada, 'Cortland' had a total winter injury of 83%, compared with 80% for 'McIntosh' and 97% for 'Early McIntosh'(9).

During the November 1940 cold period, when many fruit plants were severely damaged in the Plains States, 'Cortland' trees escaped without injury in Illinois (33). In Kansas, where 200 apple cultivars were evaluated for injury (46), trees of 'Cortland' were less hardy than 'Milton,' but showed 70 to 90% recovery. Following the severe 1947-48 winter in Minnesota (11), bud and spur killing was 29% on 'Cortland,' compared with 61% on 'McIntosh.'

After a severe freeze in British Columbia in 1966, fruit spurs of 'Cortland' were classified as hardy, while those of 'McIntosh' were very hardy (36). Because of its very winter hardy tree, 'Cortland' is a popular cultivar in very cold regions such as Quebec and New Brunswick. It can be grown in U.S. Department of Agriculture hardiness zones 4 through 6.

'Cortland' blossoms were markedly more resistant to cold injury than those of 'Delicious' or 'Wealthy' in one study (49).

Pollination

'Cortland' blooms in midseason (30, 56) with 'Delicious.' It is diploid (34 chromosomes) and its pollen is viable (63).

Knowlton (35) found that 'Cortland' produced smaller quantities of pollen (1,900 pollen grains per anther) than 11 other cultivars. 'Delicious' was the most prolific with 9,675 grains. It cannot be concluded from this, however, that 'Cortland' is a poor pollen donor.

MacDaniels and Burrell (41) reported that 'Cortland' pollen resulted in satisfactory fruit set, but was less effective than 'Delicious' pollen. In other tests, 'Cortland' lacked the capacity to set fruit when self-pollinated but was interfertile with 'McIntosh;' 75% of 'Cortland' pollen germinated, compared with 89% of 'Delicious' and 9% of 'Baldwin' pollen (42).

Using 'Cortland' pollen, MacDaniels and Heinicke (42) found the following percentages of commercial crop were set, a rating of 100% being a good, heavy commercial crop of fruit: 'Cortland' x 'Cortland,' 5%; 'Cortland' x 'Delicious,' 130%; 'Rhode Island Greening' x 'Cortland,' 150%; and 'McIntosh' x 'Cortland,' 137%. Thus, 'Cortland' pollen on 'McIntosh' gave satisfactory fruit set and resulted in an average of 5.3 seeds per fruit (14).

Although 'Cortland' is self-unfruitful, it is cross-compatible with its parent, 'McIntosh', and its half sibs, 'Milton' and 'Macoun' (37). Self-incompatibility in apples is almost universal, but one of the few known cases of pollen cross-incompatibility between two apple cultivars is 'Cortland' x 'Early Mcintosh' and the reciprocal (62). In this study, 'Cortland' was again found to be cross-compatible with 'McIntosh,' 'Melba,' 'Milton,' and 'Macoun,' all related to 'Cortland.'

'Cortland' trees are precocious and annual in bloom, making them good pollinizers for many cultivars. In New England, 'Cortland' is used mainly as a pollinizer for 'McIntosh' and 'Delicious' (47).

Oversetting of 'Cortland' blossoms can be prevented by applying thinning sprays within three weeks after full bloom using 35 to 50 parts per million of napthalacetamide (23). Today, napthaleneacetic acid is more commonly used than naphthalacetamide.

Fruit

The average ripening date for 'Cortland' at Geneva, New York, is October 7 or three days earlier than 'Delicious.' The average from full bloom to harvest is 142 days.

Unlike 'McIntosh' fruits, which drop severely just before they ripen, few 'Cortland' fruits abscise at maturity, even after they become overripe (40). Most commercial fruit growers who grow 'Cortland' also grow 'McIntosh.' Since the two cultivars ripen at about the same time, they should be harvested at the same time. However, because 'McIntosh' drops and 'Cortland' does not, growers tend to harvest 'McIntosh' before 'Cortland.' Thus, 'Cortland' fruits are often not harvested until overripe, and therefore do not store well. Continued marketing of large volumes of overripe 'Cortland' fruits for many years has resulted in the decline in popularity of this excellent cultivar.

The physical characteristics of 'Cortland' fruits, including fruit size, shape, skin, flesh, core, and others, have been described in detail (61). Dayton (19) found that the epidermal layer of 'Cortland' fruit skins has 60 to 95% of its cells pigmented. He also found that 99% of the cells in the outer hypodermal layer are pigmented (20).

The color requirements for 'Cortland' fruits in U.S. grades are U.S. Extra Fancy, 50% red; U.S. Fancy, 33%; and U.S. No. 1, 25% (38).

The vitamin C content of 'Cortland' is intermediate between that of its two parents, 'Ben Davis' and 'McIntosh' (10). In one test (31), 'Cortland' contained 11 milligrams ascorbic acid per 100 grams fruit, compared with 37 for 'Calville Blanc' and four for 'McIntosh'.

Continued marketing of large volumes of overripe 'Cortland' fruits for many years has resulted in the decline in popularity of this excellent cultivar.

'Cortland' flesh surface does not turn to a light brown color when it is exposed to air; it remains snow white.

Productivity

'Cortland' is one of the most consistently productive of all important apple cultivars. In a comparative yield test on standard rootstocks in the Hudson Valley, New York (15), the average annual yield per tree in the 16th through the 20th year was 7.3 bushels, compared with 6.8 for 'McIntosh' and 2.3 for 'Wealthy,' a biennially cropping cultivar.

In Ohio, yields of 31 apple cultivars on two types of tree frames were measured for 21 years and calculated as total yield per inch of trunk circumference (32). 'Cortland' ranked sixth in productivity, surpassed only by 'Melrose,' 'Turley,' 'Staymared,' 'Red McIntosh,' and 'Golden Delicious.'

However, 'Cortland' fails to crop well in the United Kingdom (13). Nectria canker on trunks and limbs is a problem there. This illustrates how some cultivars are better suited for some areas than others.

Storage

'Cortland' fruits keep rather well. However, bitter pit sometimes develops during storage. Storage life is about 120 days when held at 31°F. This can be greatly lengthened in controlled atmosphere (CA) storage, using 5% carbon dioxide and 3% oxygen (54). Lower oxygen levels (1.5%) are being used today for CA storage of many cultivars. Scald and sometimes carbon dioxide injury can occur in CA storage. Scald on 'Cortland' apples can be controlled by postharvest dips of 500 ppm diphenylamine (53). Before DPA became available, 'Cortland' could not be stored in CA because of scald. Vapors from other ripe apples blown over preclimacteric 'Cortland' apples at 33°F did not stimulate their respiration or rate of softening (55).

Fruit processing

'Cortland' is unique among apple cultivars in that its flesh surface does not turn to a light brown color when it is exposed to air; it remains snow white. Thus, it is excellent for the making of apple salads. It is recommended by home economists for culinary and dessert purposes (22).

However, it rates low for processing quality (3). Nevertheless, many thousands of bushels are processed annually in New York because it is an inexpensive cultivar to produce and growers bargain with processors to purchase 'Cortlands' along with high quality processing cultivars.

Diseases

'Cortland' fruits and leaves are susceptible to apple scab. Leaves are also highly susceptible to powdery mildew, especially in areas with mild winters. In a German study of this disease (50), 'Cortland' was severely attacked, while 'Lodi' was resistant. 'Cortland' is more susceptible to fireblight than 'McIntosh' but less susceptible than 'Rhode Island Greening' (29).

Crowell (16) compiled reports of the relative susceptibility of many apple cultivars to cedar-apple rust, as grown at several experiment stations. 'Cortland' was slightly susceptible.

'Cortland' trees are resistant to collar rot (6) and are also resistant to the oak wilt fungus (7).

Symptoms of the dapple apple virus have been observed on 'Cortland' (43).

Genetic mutations

A dark red sport of 'Cortland' was induced by Bishop (8) by treating dormant scions with thermal neutron radiation. It was named 'Red Cortland' and is sometimes called 'Nova Red Cortland' (48). It is not commercially acceptable because many fruits have deep sutures on one side.

'Redcort', a red-fruited mutation discovered in the Hudson Valley, New York, in 1983 (5), is patented and is being planted in commercial orchards, to the exclusion of standard 'Cortland.' Its fruits are similar to 'Cortland,' but more of the skin is colored and the shade is much darker red.

'Spur Cortland,' a spur-type mutation, was discovered by a grower in Orleans County, New York, in 1974. However, its fruits are distorted in shape and trees are unproductive, thus it is not commercially useful.

Breeding behavior

In breeding experiments at Ottawa (57), 'Cortland' was a valuable parent for large fruit size, low acidity, yellow flesh, yellow ground color, and for fruit ripening in late October. 'Cortland' was only a moderately good parent for good overcolor, quality, and late keeping. In inheritance studies (34), 'Cortland' tended to transmit large fruit size, oblate shape, and mostly fair to good quality. In other breeding studies (17, 18), 'Sandow' x 'Cortland' seedlings were promising for late keeping.

In a Swedish breeding program (45), the cross, 'Zuccalmaglios Reinette' x 'Cortland,' produced promising seedlings which were late ripening with good keeping quality, scab immunity, and cold hardiness.

Commercial usefulness

Nursery trees of 'Cortland' were sold by the New York State Fruit Testing Association, Geneva, New York, from 1921 to the early 1990s. In the 1960-68 period, 'Cortland' trees were sold by 57 nurseries—more commercial nurseries than any other Geneva apple introduction during this period. This is evidence of its widespread acceptance.

The average annual production of 'Cortland' in the United States in the 1947-50 period was 2,500,000 bushels, rising to 3,800,000 in 1962-65. The National Apple Institute ranked 'Cortland' as the tenth most important U.S. apple cultivar in 1964 (4) and 11th in 1966 (21). In 1995, 80 million pounds (nearly 2 million bushels) of 'Cortland' were produced in New York (51), making it the sixth most important apple cultivar in the state. In descending order of importance, the other five were 'McIntosh,' 'Rome,' 'Delicious,' 'Rhode Island Greening,' and 'Empire.' 'Cortland' production is regional, with most produced in the Northeast. According to E.C. Wilcox, the 1952 'Cortland' crop was 75% from New York and Pennsylvania and 13% from New England; Ohio and California were other important producers that year (59).

Compared with 'McIntosh,' 'Cortland' fruits ripen a little later, hang on the tree much more tenaciously, keep longer in storage, bruise less in handling, are larger, and are brighter in color (26). Because of these outstanding attributes, 'Cortland' is still recommended for planting in New York and New England and is still popular with growers and consumers over 80 years after its introduction. It is a variety that has withstood the test of time.

Literature cited

1) Anonymous. 1934. N.Y. State Agr. Exp. Sta. Rpt. 53: 53-65.

2) Anonymous. 1935. N.Y. State Agr. Exp. Sta. Rpt. 54: 62-80.

3) Anonymous. 1954. Apple varieties for freezing. *Fruit Var. Hort. Dig.* 9:37.

4) Anonymous. 1964. National Apple Institute June 1964 crop guesstimate by varieties. *Nat. Apple News* 4(7):6.

5) Anonymous. 1983. 'Redcort.' *Fruit Grower,* April, p. 62.

6) Baines, R.C. 1939. Phytophthora trunk canker or collar rot of apple trees. *J. Agr. Res.* 59: 159-184.

7) Bart, G.J. 1960. Susceptibility of various apple varieties to the oak wilt fungus. *Phytopathology* 50:177-178.

8) Bishop, C.J. 1959. Color sports of apples induced by radiation. *Fruit Var. Hort. Dig.* 14:37-39.

9) Blair, D.S. 1935. Winter injury to apple trees in eastern Canada 1933-35. *Sci. Agr.* 16:8-15.

10) Bregger, M.P. and J.T. Bregger. 1937. Vitamin and other nutritional values in apple varieties. *Virginia Fruit* 25: 147-160.

11) Brierly, W.G., W.H. Alderman, and T.S. Weir. 1950. Winter injury to apple trees in Minnesota 1947-48. *Proc. Amer. Soc. Hort. Sci.* 55:259-261.

12) Brooks, R.M. and H.P. Olmo. 1952. *Register of New Fruit and Nut Varieties 1920-50.* Univ. Calif. Press.

13) Bultitude, J. 1983. *Apples.* Univ. Wash. Press, Seattle.

14) Burrell, A.B. and R.G. Parker. 1931. Pollination of the 'McIntosh' apple in the Champlain Valley. *Proc. Amer. Soc. Hort. Sci.* 28:78-84.

15) Cole, A.B. 1947. Yields of apple varieties. *Fruit Var. Hort. Dig.* 2:101-103.

16) Crowell, I.H. 1934. Compilation of reports on the relative susceptibility of orchard varieties of apples to the cedar-apple rust disease. *Proc. Amer. Soc. Hort. Sci.* 32:261-272.

17) Davis, M.B., D.S. Blair, and L.P.S. Spangelo. 1954. Apple breeding at the Central Experimental Farm, Ottawa, Canada 1920-51. *Proc. Amer Soc. Hort. Sci.* 63:243-250.

18) Davis, M.B., D.S. Blair, and L.P.S. Spangelo. 1956. Breeding for early and late maturity. Prog. Rpt. Can. Dept. Agr., Hort. Div. 1949-53: 17.

19) Dayton, D.F. 1959. Red color distribution in apple skin. *Proc. Amer. Soc. Hort. Sci.* 74:72-81.

20) Dayton D.F. 1963. The distribution of red color in the skin of apple varieties of 'McIntosh' parentage. *Proc. Amer. Soc. Hort. Sci.* 82:51-55.

21) Dominick, B.A. 1966. National apple production trends. Proc. Nat. Apple Ind. Utilization Conf., Univ. Md., March 23-25.

22) Dunn, M. 1963. Apples, a favorite food. Cornell Ext. Bull. 973.

23) Forshey, C.G. and M.B. Hoffman. 1967. Factors affecting chemical thinning of apples. N.Y. State Agr. Exp. Sta. Res. Circ. 4.

24) Hedrick, U.P. 1915. Second distribution of Station apples. N.Y. State Agr. Exp. Sta. Circ. 37.

25) Hedrick U.P. 1923. New or noteworthy fruits VI. N.Y. State Agr. Exp. Sta. Bull. 497.

26) Hedrick, U.P. 1925. New or noteworthy fruits VIII. N.Y. State Agr. Exp. Sta. Bull. 531.

27) Hedrick, U.P. 1926. Some new or noteworthy fruits. N.Y. State Agr. Exp. Sta. Circ. 83.

28) Hedrick, U.P. and R. Wellington. 1912. An experiment in breeding apples. N.Y. State Agr. Exp. Sta. Bull. 350.

29) Hildebrand, E.M. and A.J. Heinicke. 1937. Incidence of fireblight in young apple trees in relation to orchard practices. Mem. Cornell Agr. Exp. Sta. 203.

30) Hoffman, M.B. 1965. Pollination and fruit development of tree fruits. Cornell Agr. Exp. Sta. Bull. 1146.

31) Howe, G.H. and W.B. Robinson. 1946. Ascorbic acid content of apple varieties and seedlings at Geneva, N.Y. in 1944-45. *Proc. Amer. Soc. Hort. Sci.* 48:133-136.

Authors Way and Brown were with the Department of Horticultural Sciences, New York State Agricultural Experiment Station, Cornell University, Geneva, New York, when this article originally appeared, April 1997.

32) Howlett F.S. 1969. Total yield and tree circumference of 31 apple cultivars on their own roots and on Hibernal framework, 1948-68. Ohio Agr. Res. and Dev. Cen. Res. Summary 36:45-47.

33) Kelley, V.W. and R.L. McMunn. 1942. November 1940 cold damage to fruit plants in Illinois. *Proc. Amer. Soc. Hort. Sci.* 40:220-224.

34) Klein, L.G. 1958. The inheritance of certain fruit characters in the apple. *Proc. Amer. Soc. Hort. Sci.* 72: 1-14.

35) Knowlton, H.E. 1936. The relative abundance of pollen production of apples. *Proc. Amer. Soc. Hort. Sci.* 32:7-9.

36) Lapins, K. 1967. Winter hardiness of apple varieties. *Fruit Var. Hort. Dig.* 21:78-79.

37) Latimer, L.P. 1937. Self- and cross-pollination in the 'McIntosh' apple and some of its hybrids. *Proc. Amer. Soc. Hort. Sci.* 34:19-21.

38) Lennartson, R.W. 1963. United States standards for grades of apples. USDA Fed. Reg. Doc. 63-9518.

39) MacDaniels, L.H. 1925. Pollination studies with certain New York State apple varieties. *Proc. Amer. Soc. Hort. Sci.* 22:87-96.

40) MacDaniels, L.H. 1937. Some anatomical aspects of apple flower and fruit abscision. *Proc. Amer. Soc. Hort. Sci.* 34:122-129.

41) MacDaniels, L.H. and A.B. Burrell. 1932. The value of the pollen of some of the recently introduced apple varieties. *Proc. Amer Soc. Hort. Sci.* 29:151-155.

42) MacDaniels, L.H. and A.J. Heinicke. 1929. Pollination and other factors affecting the set of fruit with special reference to the apple. Cornell Agr. Exp. Sta. Bull. 497.

43) McCrum, R.C., J.G. Barrat, M.T. Hilborn, and A.E. Rich. 1960. An illustrated review of apple virus diseases. Maine Agr. Exp. Sta. Bull. 595.

44) Nitschke, R.A. 1965. Apples for dessert, a second look (Part II). *Fruit Var. Hort. Dig.* 9:63-66.

45) Nybom, N. and P.O. Bergendal. 1960. (Pome fruits and bush fruits.) (Swedish) Rpt. Balsgård Fruit Breed. Inst. for 1939: 1-13 (From Plant Breed. Abst. 31:463).

46) Pickett, B.S. and H.L. Lantz. 1942. Apple varieties: behavior of two hundred varieties following the freeze of November 1940. *Proc. Amer. Soc. Hort. Sci.* 40:212-214.

47) Rasmussen, E.J. 1958. Apples in New England. *Fruit Var. Hort. Dig.* 12:61.

48) Ritter, C.M. 1968. Sources of scionwood of apple cultivars, rootstocks, and species in agricultural experiment stations of the United States and Canada. *Fruit Var. Hort. Dig.* 22:23-39.

49) Roberts, R.H. 1946. Cold injury of apple blossoms, 1945. *Proc. Amer Soc. Hort. Sci.* 47:61-63.

50) Schander, H. 1958. (Investigations on the development of early selection methods for apple breeding. II. Early selection for resistance to powdery mildew (*P. leucotricha* Salm.), the susceptibility of apple varieties and the inheritance of susceptibility.) (German) Zuchter 28:105-132.

51) Schooley, R.E. and R.D. Breyo. 1996. 1995 annual summary of New York fruit production. N.Y. Agr. Statistics Serv., N.Y. Dept. Agr. and Mkts., Albany, NY.

52) Shaw, J.K. 1943. Descriptions of apple varieties. Mass. Agr. Exp. Sta. Bull. 403.

53) Smock, R.M . 1957. A comparison of treatments for control of the apple scald disease. *Proc. Amer. Soc. Hort. Sci.* 69:91-100.

54) Smock, R.M. 1958. Controlled-atmosphere storage of apples. Cornell Agr. Expt. Sta. Ext. Bull. 759.

55) Smock, R.M. and L. Yatsu. 1959. Apple volatile effects on the ripening of 'Cortland' apples. *Proc. Amer. Soc. Hort. Sci.* 73:71-77.

56) Southwick, F.W. 1963. Pollination of fruit plants. Mass. Agr. Exp. Sta. Spec. Circ. 247.

57) Spangelo, L.P.S. 1960. Fruit breeding. Prog. Rpt. Hort. Div. Central Exp. Farm, Ottawa. 1954-58:21-24.

58) Tukey, H.B. 1976. History of the American Pomological Society. *In:* Fisher, D.V. and W.H. Upshall. *History of fruit growing and handling, 1860-1972.* Regatta City Press, Kelowna, British Columbia.

59) Upshall, W.H. 1970. *North American apples: varieties, rootstocks, outlook.* Mich. State Univ. Press, East Lansing, Michigan.

60) Way, R.D. 1967. An expanded apple breeding program at Geneva. Farm Res. 33(4):8-9.

61) Way, R.D. 1971. Apple cultivars introduced by the New York State Agricultural Experiment Station, 1914 to 1968. N.Y. State Agr. Exp. Sta. Search 1 (2).

62) Weeks, W.D. and L.P. Latimer. 1939. Incompatibility in 'Early McIntosh' and 'Cortland' apples. *Proc. Amer. Soc. Hort. Sci.* 36:284-286.

63) Wellington, R. 1947. Pollination of fruit trees. Cornell Agr. Exp. Sta. Ext. Bull. 720.

The 'Marion' Trailing Blackberry

by Chad Finn, Bernadine Strik, & Francis J. Lawrence

'Marion' is the most widely planted blackberry cultivar in the world. In Oregon, which leads the world in production of blackberries, 'Marion,' often called "marionberry" by consumers and marketers, has been the dominant cultivar since the early 1980s when it replaced 'Thornless Evergreen' as the most widely planted cultivar (15). In 1995, 'Marion' was harvested from 1,420 hectares within Oregon. Approximately 200 hectare was in the "off-year" of alternate year production (16) in 1995.

About 95% of the Pacific Northwest blackberry crop is processed (16). 'Marion' and 'Thornless Evergreen' account for approximately 70% and 20% of the hectarage, respectively (15). 'Marion' is regarded as a berry with a premium quality and is usually sold under the 'Marion' name, whereas 'Thornless Evergreen' and other blackberries are sold under a generic, "blackberry" label.

'Marion' has developed its outstanding reputation for several reasons, primarily related to fruit quality, including fruit flavor, aroma, and perception of fewer pyrenes. George E. Waldo (23), the USDA-ARS small fruit breeder in Corvallis, Oregon, was able to incorporate the outstanding flavor and pleasant aroma of the trailing western blackberry (*Rubus ursinus* Cham. & Schitdl.; syn. *R. macropetalus* Doug.) into 'Marion.' 'Marion' is perceived as being "less seedy" than 'Thornless Evergreen,' eastern U.S. erect, and semi-erect blackberry cultivars. While pyrene measurements have not identified any size differences between 'Marion' and eastern U.S. cultivars (Takeda, pers. comm.), there have been several proposed reasons for this perceived difference. 'Marion' pyrenes have a different shape, they are flatter than the ellipsoidal and clam-shaped eastern cultivars, thus sliding more easily between your teeth (20). 'Marion' has a soft, thin endocarp in comparison to the eastern cultivars (20); also, the pyrenes are coated with a 'gelatinous' material so that there is a "cushioning" effect when the fruit is eaten.

'Marion' represents an amazingly diverse ancestry.

'Marion,' a hexaploid, was released in 1956 by the cooperative breeding program of the U.S. Department of Agriculture-Agricultural Research Service and the Oregon Agricultural Experiment Station. The name 'Marion' was chosen to recognize Marion County, Oregon, where the berry was tested extensively. In 1948, 'Marion' was selected by Waldo from a cross of 'Chehalem' x 'Olallie' made in 1945 (23). The pedigree of 'Marion' is quite diverse, leading consumers to feel unsure as to what type of blackberry is 'Marion' (Fig. 1).

The pedigree of 'Marion' is also confusing to scientists. Based on updated taxonomy of early reports (5, 6, 11, 12, 13, 14, 21, 23, 24), the pedigree contains *R. ursinus* (approximately 44%), *R. armeniacus* Focke (25%; syn. *R. procerus* Muller), *A. flagellaris* Willd. (13%), *R. aboriginum* (13%) arid *R. idaeus* L. (6%). However, 'Marion's pedigree appears to be more complicated than this, and it may never be determined with complete accuracy.

For instance, 'Santiam,' a chance seedling found by a grower, is perfect flowered while native *R. ursinus* is dioecious. Although 'Santiam' appears to be largely derived from *R. ursinus*, the perfect flowering characteristic may have come from 'Logan,' which was commercially grown at the time 'Santiam' was found. The cultivar 'Black Logan' also has an uncertain origin as does 'Phenomenal;' in these two cases the maternal parent is known, but the pollen parent has been hypothesized based on the different *Rubus* species and cultivars growing in the vicinity. While we cannot be positive about the entire ancestry of 'Marion,' the fruit characteristics and plant growth habit are most similar to *R. ursinus*.

Trailing blackberry cultivars tend to be extremely vigorous, and 'Marion' is no exception. New primocanes emerge in the spring, and grow upright until cane weight pulls them to the ground where they grow along the soil surface. Primocanes can grow 6 to 11 meters during the growing season (2, 4), with one plant producing as much as 200 meters of primocane growth (4). Plants also produce a second flush of primocanes, naturally, during the fruit ripening period (4).

'Marion's' poor winter tolerance is the main factor that has limited its production to the Willamette Valley. Depending on the environmental conditions leading up to and after a winter cold period, state of dormancy, and cultural practices, 'Marion' often will exhibit cane and/or bud injury (LT_{50}) at -5 to -22°C (23° to -8°F) (3, 4). Apparently, 'Marion' has a very low chilling requirement that is often met before winter has even begun, making it particularly susceptible to fluctuating winter temperatures (17). The plants exhibit a remarkable ability to develop secondary buds and will often produce a full crop on secondary buds after the primary buds have been killed (18).

Growers typically bundle and tie the primocanes to a trellis in mid-August to early September; occasionally, they are left on the ground until February before being trained. The timing of the trellising reflects the risk a grower is willing to take relative to potential winter injury. August-September trained 'Marion' plantings are more productive the following year, (2) but since they are on the trellis as opposed to the ground during the winter, they are more susceptible to winter injury (1). Primocane suppression date can also affect subsequent cold hardiness and yield (2, 3).

During the fruiting year, the plants break bud in early spring but do not normally flower until after the danger of frost has passed. The crop typically begins to ripen at the end of June, with commercial harvest beginning the first week of July and finishing in late July. Over 85% of the crop is harvested with mechanized harvesters (16). Fruit for the IQF (individually quick frozen) market and the very limited fresh market is usually harvested by hand; the lack of fruit firmness is a major limitation for 'Marion.' The drupelet skin will often break under the weight of other fruit in the harvest flat, thus hindering 'Marion's' use for fresh market. More importantly, it is difficult in the processing plants to use air blowers to separate leaves and other contaminants from the fruit when the fruit are compacted and have leaked juices from their broken skins.

'Marion' canes have small prickles. These are a nuisance during training, but they are a serious problem when these prickles dislodge during mechanical harvesting. The prickles become a contaminant in the harvested fruit.

'Marion' fruit, which average 4.5 to 5.5 grams, typically have about 65 to 80 drupelets per fruit early in the season and 60 to 70 later in the season (7, 8, 9, 19). Ripe fruit retain their color well when processed, and the fruit averages 13.6% soluble solids and 1.5% titratable acidity, with a pH near 3.2.

Fruit yield of 'Marion' is not particularly high. In replicated trials, 'Marion' plants produced

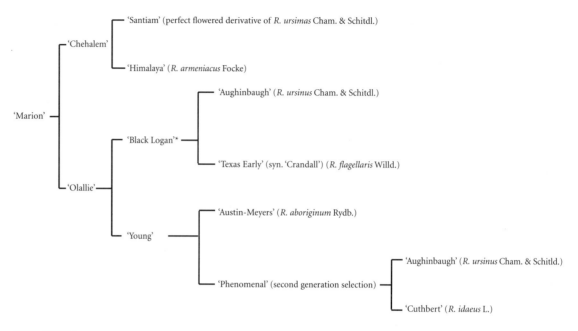

FIGURE 1.

The pedigree of 'Marion' trailing blackberry.

* 'Black Logan' is either 1) synonymous with 'Mammoth,' and it is a second-generation selection from 'Aughinbaugh' x 'Texas Early' (12, 13); or 2) it is a selection from open-pollinated seed of 'Mammoth' that in turn was a selection from 'Aughinbaugh' x 'Texas Early' (6). Information derived in part from: Darrow 1918; 1937; Hedrick, 1925 Hall, 1990; Jennings, 1981; 1988; Logan, 1909; Thompson, 1997; Waldo, 1957; Waldo and Darrow, 1948.

The outstanding fruit quality, particularly flavor, of 'Marion' has been the reason it has risen to such dominance in the worldwide market.

15,000 kg/ha in their second cropping season (7, 8, 9). In commercial production, the average yield for 1993-95 was 7,500 kg/ha; however, some cold injury occurred in these years. In comparison, 'Thornless Evergreen,' a more cold-hardy derivative of *R. Iaciniatus* (1), averaged 9,600 kg/ha over this same period, and growers report yields of 'Chester Thornless,' a semi-erect blackberry, at 22,500 kg/ha.

While 'Marion' is susceptible to cane and foliar diseases, such as septoria leaf spot (*Septoria rubi* Westend) and purple blotch (*Septocyta ruborum* (Lib.) Petr.), and fruit pests, such as botrytis fruit rot, dryberry mite (*Phyllocoptes gracilis* [Nalepa]), and redberry mite (*Acalitus essigi* [Hassan]), these pests can be controlled with good management. As with most blackberries, 'Marion' is tolerant of root diseases. While the trailing blackberries can be infected with tobacco streak virus (TSV), raspberry bushy dwarf virus BDV), and blackberry calico virus (BCV), these viruses have not been shown to affect growth or yield in 'Marion.'

'Marion' represents an amazingly diverse ancestry. The breeders of the past were able to capture several of the most positive characteristics of a species and combine it into one genotype. The outstanding fruit quality, particularly flavor, of 'Marion' has been the reason it has risen to such dominance in the worldwide market. 'Marion' will continue to be the predominant cultivar in the Pacific Northwest until a cultivar is developed that is firmer fruited, more winter cold tolerant, thornless, and, most importantly, retains 'Marion's' fruit quality characteristics.

Literature cited

1) Bell, N., E. Nelson, B. Strik, and L. Martin. 1992. Assessment of winter injury to berry crops in Oregon, 1991. Agr. Exp. Sta. Spec. Rpt. 902. Oregon State University.

2) Bell, N.C., B.C. Strik, and L.W. Martin. 1995a. Effect of primocane suppression date on 'Marion' trailing blackberry. I. Yield, components. *J. Amer. Soc. Hort. Sci.* 120:21-24.

3) Bell, N.C., B.C. Strik, and L.W. Martin. 1995b. Effect of primocane suppression date on 'Marion' trailing blackberry. I. Cold hardiness. *J. Amer. Soc. Hort. Sci.* 120:25-27.

4) Cortell, J. and B. Strik. 1997. Effect of floricane number in 'Marion' trailing blackberry. I. Primocane growth and cold hardiness. *J. Amer. Soc. Hort. Sci.* (in press).

5) Darrow, G.M. 1918. Culture of Logan Blackberry and Related Varieties. USDA Farmers' Bulletin 998. Washington, D.C. (24 pp).

6) Darrow, G.M. 1937. Blackberry and raspberry improvement. pp. 496-533. *In:* Better plants and animals-II. *USDA Yearbook of Agriculture.* U.S. Government Printing Office, Washington, D.C.

7) Finn, C., K. Wennstrom, T. Mackey, D. Peacock, and G. Koskela. 1996. New small fruit cultivars and advanced selections for the Pacific Northwest. *Proc. Ore. Hort. Soc.* 87:117-120.

8) Finn, C. 1997. New small fruit cultivars from down South; USDA releases from Corvallis. *Proc. W. Wash. Hort. Assoc.* 141-144.

9) Finn, C. and F.J .Lawrence. 1997. 'Black Butte' trailing blackberry. *HortScience* (submitted).

10). H.K. 1990. Blackberry breeding. pp. 249-302. *In:* J. Janick (ed.) *Plant Breeding Reviews,* Vol. 8. Timber Press, Portland, Oregon.

11) Hedrick, U.P. 1925. The Small Fruit of New York. State Department of New York-Farms and Markets. *Thirty-third Annual Report-Part II,* J.B. Lyon Co., Albany, NY.

12) Jennings, D.L. 1981. A hundred years of loganberries. *Fruit Var. J.* 35(2):34-37.

13) Jennings, D.L. 1988. Raspberries and Blackberries: Their breeding, diseases, and growth. Academic Press, New York.

14) Logan, J.H. 1909. Loganberry, Logan Blackberry, and Mammoth Blackberry. The Pacific Rural Press and California Fruit Bulletin. September 25, 1909. Vol. 78:193, 196.

15) Oregon Agricultural Statistics Service. 1981-1996. Berry Crop Summary. Portland, Oregon.

16) Strik, B.C. 1992. Blackberry cultivars and production trends in the Pacific Northwest. *Fruit Var. J.* 46:202-206.

17) Strik, B., H. Cahn, N. Bell, and J. deFrancesco. 1994. Caneberry research at North Willamette Research and Extension Center-an update. *Proc. Ore. Hort. Soc.* 85:141-149.

18) Strik, B., H. Cahn, N. Bell, J. Cortell, and J. Mann. 1996a. What we've learned about 'Marion' blackberry—potential alternative production systems. *Proc. Ore. Hort. Soc.* 87:131-136.

19) Strik, B., J. Mann, and C. Finn. 1996b. Percent drupelet set varies among blackberry genotypes. *J. Amer. Soc. Hort. Sci* 121:371-373.

20) Takeda, F. 1993. Characterization of blackberry pyrenes. *HortScience* 28:128 (Abstract).

21) Thompson, M.M. 1997. Survey of chromosome numbers in *Rubus* (Rosaceae: Rosoideae); *Ann. Missouri Bot. Gard.* 84:128-164.

22) Waldo, G.E. 1977. 'Thornless Evergreen'—Oregon's leading blackberry. *Fruit Var. J.* 31:26-30.

23) Waldo, G.E. 1957. The Marion Blackberry. Oregon State College, Corvallis, Ore. Circular of Information 571, 7 pages.

24) Waldo, G. and G.M. Darrow. 1948. The origin of the Logan and Mammoth blackberries. *J. Hered.* 39:99-107.

Authors Finn and Strik were with the USDA-ARS Horticultural Research Lab, Corvallis, Oregon; and the Department of Horticulture, Oregon State University, Corvallis, Oregon, respectively; and author Lawrence was a retired horticulturist in Corvallis, Oregon, when this article originally appeared, July 1997.

The 'Sunred' Nectarine

by *Jeffrey G. Williamson & Wayne B. Sherman*

'Sunred' nectarine is a product of the University of Florida's *Prunus* breeding program. A brief recounting of the early history of this program is provided. Professor Ralph Sharpe began breeding low-chill peaches at the University of Florida in 1952 after field tests in Florida revealed that none of the peach and nectarine cultivars available at the time were suitable for commercial production in Florida (11). Sharpe's primary goal was to develop early-season, low-chill peach cultivars with good commercial qualities that would ripen in central Florida during April and May (before California shipments began).

The first release came in 1961 when 'Flordawon' was named and suggested for trial in central Florida (5). Its main attributes were low chill requirement and early harvest date (late April or early May). It is well adapted to central Florida, but produces fruit that are too soft for commercial shipping and handling (7). Commercial plantings of 'Flordawon' never exceeded 25 acres.

More suitable cultivars such as 'Early Amber' and 'Flordasun' peaches, and 'Sunred' nectarine, quickly became available in the mid 1960s and established themselves as major cultivars for the central Florida industry. (1). 'Early Amber,' a patented cultivar, was developed by Peaches of Florida, Inc., using one of the University of Florida's breeding lines. 'Flordasun' and 'Sunred' were released by the University of Florida in 1964. Together, these three cultivars represented over 85% of the central Florida peach and nectarine acreage during the late 1960s and early 1970s (1, 2).

The nectarine character was first introduced into the breeding program in 1956 by growing an F2 population from the cross, 'Panamint' nectarine x ('Southland' x 'Hawaiian') F2 (5, 6). 'Sunred' nectarine was selected from this cross in 1961 and observed in Gainesville and Deland, Florida, before its release in 1964 (6). In central Florida, 'Sunred' trees reliably produce bright red, roundish, semi-freestone fruit with firm, yellow flesh of excellent dessert quality. With over 30 years of experience with 'Sunred' in Florida, it has become our standard for 250 chill units and is used to rate other cultivars for chill requirement. 'Sunred' usually blooms about February 1 in central Florida. The average first picking date for 'Sunred' in central Florida is May 10, which was usually about two weeks before California began shipping nectarines during the late 1960s and early 1970s. Early harvest season compensated for the small fruit size of 'Sunred,' which tended to be 1-3/4 to 2 inches in diameter.

'Sunred' quickly became established as the major commercial nectarine grown in central Florida and was one of the three most widely planted cultivars (along with 'Flordasun' and 'Early Amber' peaches) in Florida between 1967 and 1969. By 1969, 'Sunred' occupied over 750 acres of the 2,800-acre peach/nectarine industry in central Florida. The high grower interest in 'Sunred' was no doubt related to the excellent prices received for fresh nectarines during late April and early May. In 1970, Florida-grown nectarines returned profits of between 30 and 40 cents per pound after picking and packing (2).

At peak production in 1970, there were 31 car lots of 'Sunred' fruit shipped out of Florida (8). During the same period that 'Sunred' became established as the standard for early-season nectarines in central Florida, it was also being evaluated for its commercial potential in other regions of the United States, Europe, Asia, and Central and South America (9).

Eventually, 'Sunred,' with its small fruit size, could not compete with the new, earlier-ripening, cultivars which were being planted in California during the early 1970s. By the late 1970s, commercial peach acreage (including 'Sunred') in central Florida had declined significantly because of these new cultivars from California.

'Sunred' represented a major advance in early-season, low-chill nectarine cultivar development. It demonstrated that commercial production of low-chill nectarines for the early-season market was feasible in central Florida, and in certain other tropical and subtropical regions of the world.

However, 'Sunred's' greatest and longest-lasting contribution to the low-chill stone fruit industries probably comes from its continued use in breeding and cultivar improvement. 'Sunred' has been used as a breeding line in Florida to develop numerous low-chill peach and nectarine cultivars which are currently grown in several counties. For example, 'Sundowner' nectarine (Fla. 6-3N) and 'Forestgold' peach (Fla. 7-11), both named and grown in Australia, had 'Sunred' in their lineage, as did 'Maravilha' (Fla. 13-72) peach, which was named in Brazil and grown in Brazil and Australia. During the mid-1980s, 'Sunred' was still

> *Eventually, 'Sunred,' with its small fruit size, could not compete with the new, earlier-ripening cultivars which were being planted in California during the early 1970s.*

the most widely grown nectarine in the low-chill stone fruit producing areas of Australia (4). However, by 1990, 'Sundowner' had replaced 'Sunred' as the major low-chill nectarine grown there (11).

Other low-chill cultivars from Florida which resulted from using 'Sunred' in breeding include 'Flordaprince,' 'Flordaglo,' 'TropicSnow,' and 'Sunhome.' 'Sunred' has been replaced in Florida by 'Sunraycer,' which blooms and ripens with 'Sunred' but has larger fruit than 'Sunred.'

The legacy of 'Sunred' nectarine can be summarized as follows: 1) it was the first commercial nectarine cultivar to become established in low-chill production areas around the world; and 2) it was used as a parent in peach and nectarine breeding programs in Florida and California. 'Flordaprince,' a low-chill peach which is a descendent of 'Sunred' and which was recognized for its worldwide importance when it received the ASHS Fruit Breeders' Working Group Outstanding Cultivar Award, continues to be a major cultivar in many subtropical production areas of the world.

Literature cited

1) Anonymous. 1969. 5,640 acres of commercial peaches in Florida. 1969 Peach Report. Florida Crop and Livestock Reporting Service, Orlando, FL.

2) Anonymous. 1972. Florida peach and nectarine trees total over 800,000. 1972 Peach Report. Florida Crop and Livestock Reporting Service, Orlando, FL.

3) Loebel, R. 1987. Low-chill stone fruit industry around Australia. *In:* I. Skinner (ed.). *Proc. of the First National Low-chill Stonefruit Conference.* pp. 39-54.

4) Malcolm, P.J. 1991. N.S.W. Stone, berry and miscellaneous temperate fruit statistics. Agnote ISSN 1034-6848, Division of Plant Industries NSW Agriculture and Fisheries.

5) Sharpe, R.H. 1961. Developing new peach varieties for Florida. *Proc. Fla. State Hort. Soc.* 74:348-352.

6) Sharpe, R.H. 1964. 'Sunred.' Cir. S-158. Agricultural Exp. Station, University of Florida, Gainesville.

7) Sharpe, R.H. 1969. Sub-tropical peaches and nectarines. *Proc. Fla. State Hort. Soc.* 82:302-308.

8) Sharpe, R.H. and J.B. Aitken. 1971. Progress of the nectarine. *Proc. Fla. State Hort. Soc.* 84:338-345.

9) Sharpe, R.H., T.E. Webb, and H.W. Lundy. 1954. Peach variety tests. *Proc. Fla. State Hort. Soc.* 67:245-246.

10) Sherman, W.B., J. Soule, and C.P. Andrews. 1977. Distribution of Florida peaches and nectarines in the tropics and subtropics. *Fruit Var. J.* 31:75-78.

11) Sweedman, R. 1992. The low-chill stonefruit industry in NSW and Australia. *In:* J.M. Slack (ed.). *Proc. of the Second National Low-chill Stonefruit Conf.* pp. 5-16.

Authors Williamson and Sherman were with the Horticultural Sciences Department, University of Florida, Gainesville, Florida, when this article originally appeared, January 1996.

Wilder Award Recipients

Presented by the American Pomological Society, 1947-1997

by Dennis Werner

The Wilder Medal was established in 1873 by the American Pomological Society in honor of Marshall Pinckney Wilder, the founder and first president of the Society. The award consists of a beautifully engraved medal, which is usually presented to the recipient at the banquet during the annual meeting of the Society.

The Wilder Medal is conferred on individuals or organizations which have rendered outstanding service to horticulture in the broad area of pomology. Special consideration is given to work relating to the origination and introduction of meritorious varieties of fruit. Individuals associated with either commercial concerns or professional organizations may be considered as long as their introductions are truly superior, and have been widely planted.

Significant contributions to the science and practice of pomology other than fruit breeding also will be considered. Such contributions may relate to any important area of fruit production, such as rootstock development and evaluation, anatomical or morphological studies, or unusually noteworthy publications in any of the above subject areas. One or more Wilder Medals may be conferred each year, but the number is usually limited to three.

The recipients of the Wilder Medal from 1947 through 1997 are as follows:

1947	**Maurice A. Blake**, New Jersey. Peach breeder.
1948	**Liberty H. Bailey**, Cornell. Plant explorer, writer, horticulturist.
	William H. Chandler, California. Research in pomology and hort. publications.
	George M. Darrow, USDA. Fruit breeding, esp. blueberries and strawberries.
	Ezra J. Kraus, Oregon. & University of Chicago. Economic botany, morphology & physiology.
	Horticultural Experiment Station, Vineland, Canada. Origination of fruit varieties.
	Ruby grapefruit. Outstanding red-fleshed grapefruit.
1949	**Frank C. Reimer**, USDA, Oregon. Pear breeding.
	University of Minnesota. Fruit Breeding Farm. Origination of fruit varieties.
1950	**H. Howard Hume**, Florida. Subtropical and tropical horticulture.
1950	**Archibald D. Shamel**, California. Introduction of tropical and subtropical fruits, mutations.
	Jacob K. Shaw, Massachusetts. Identification of fruit tree varieties by leaf characteristics.
	Henry P. Stuckey, Georgia. Muscadine grape breeding and pecan culture.
1951	**Stanley Johnston**, Michigan. Breeding peaches and blueberries.
	Richared apple. Custody of Louis Richardson, Manitou, Washington.
	Central Experiment Farm, Ottawa, Canada. Fruit breeding research.
1952	**Eugene C. Auchter**, USDA. Pollination biology, horticultural writer, administrator.
	Elizaeth White, New Jersey. Blueberry breeding and introduction.
	Louisiana Agriculture Experiment Station. Klonmore strawberry.

1953	**Maxwell J. Dorsey**, Illinois. Peach breeding.
	Samuel Fraser, Cornell and American Apple Institute. Apple promotion, new fruit varieties.
	George F. Potter, New Hampshire and USDA. Winter hardiness, tung genetics and physiology.
	United States Department of Agriculture. Origination of meritorious varieties of peaches.
1954	**Paul H. Shepard**, Missouri. Peach, apple and grape breeding.
	Richard Wellington, New York. Fruit breeding and evalution, pollination.
	California Agriculture Experiment Station. Origination of strawberry varieties.
1955	**William H. Alderman**, Minnesota. Breeding of hardy fruits.
	Frederick W. Popenoe, USDA. Plant explorer, tropical and subtropical horticulture.
1956	**John R. Magness**, USDA. Physiology, nutrition, water relation, administration.
	Elmer Snyder, California. Grape breeding and resistant rootstocks.
	Harold B. Tukey, New York and Michigan. Morphology of fruits, size-controlling rootstocks.
1957	**W.W. Magill**, Kentucky. Outstanding extension horticulturist.
	Edmund F. Palmer, Vineland, Canada. Breeding fruits, especially peaches.
	Warren P. Tufts, California. Pollination and morphology of deciduous fruits.
	Albert F. Yeager, North Dakota and New Hampshire. Breeding and genetics of fruits.
1958	**Cecil L. Burkholder**, Indiana. Variety evaluation, pollination.
	Victor R. Gardner, Oregon and Michigan. Pollination, fruit set and mutation studies.
	Henry Hartman, Oregon. Postharvest physiology, esp. of pears.
	Ralph A. Van Meter, Massachusetts. Systematic pomologist and administration.
1959	**Malcolm B. Davis**, Ottawa, Canada. Nutrition and low temperature research.
	Harold P. Olmo, California. Grape breeding, cytogenetics. New Fruit Register.
	Thomas J. Talbert, Missouri. Fruit tree rootstocks and nutrition.
	New York Fruit Testing Association, Geneva, New York. Promotion of new fruit varieties.
1960	**Arthur P. French**, Massachusetts. Nursery identification of fruit trees by leaf characters.
1961	**Richard P. White**, Am. Assoc. Nurserymen, Wash., D.C. Horticultural promotion.
1962	**Robert W. Hodgson**, California. Subtropical and tropical horticulture., international service.
	Leif Verner, Idaho. Posthumous. Breeding, training, thinning, water relations.
	United States Department of Agriculture. Development of virus-free strawberries.
1963	No recipients.
1964	**Frank P. Cullinan**, USDA. Administration, peach breeding and physiology.
1965	No recipients.
1966	**Laurence H. MacDaniel**, Cornell. Pollination, fruit anatomy and morphology.
	Howard B. Frost, California. Citrus breeding and cytogenetics.
1967	No recipients.
1968	**Reid M. Brooks**, California. Posthumous. Flower and fruit morphology. New Fruit Register.
	George L. Slate, New York. Breeding strawberry, raspberry, blackberry.
1969	**Fred W. Anderson**, California. Breeding nectarines, peach, plum, almond.
	John H. Weinberger, California. Breeding peach, apricot, plum.
1970	**Joseph R. Furr**, California. Breeding and water relations of citrus.
	George Oberle, Virginia. Breeding peach, nectarine, apple, grape.
	George F. Waldo, Oregon. Breeding strawberry, blackberry and red raspberry.
1971	No recipients.
1972	**Paul Stark**, Sr. Missouri. Nurseryman, development of dwarfing rootstocks.
1973	**O.A. Bradt**, Ontario. Breeding peaches and apricots.
1974	**Robert F. Carlson**, Michigan. Breeding apple rootstocks.
1975	**Chester D. Schwartze**, Oregon. *Rubus* breeding.
	Ralph H. Sharpe, Florida. Peach breeding.
1976	**Franklin Correll**, North Carolina. Peach breeding.
1977	**Donald H. Scott**, Maryland. Strawberry breeding.
	W.A. Luce, Washington. Fruit agent, orchard consultant and Contributing editor to *The Goodfruit Grower* magazine.
1978	**L. Fredrick Hough**, New Jersey. Peach, apple and pear breeding.
	Harold W. Fogle, Maryland. Peach, apricot and cherry breeding.
1979	**Royce S. Bringhurst**, California. Strawberry breeding.
1980	**Melvin N. Westwood**, Oregon. Teacher, author and researcher of apples and pears.

1981	**Frank Street**, Kentucky. Peach grower and host of annual peach pruning demonstrations.	1991	**Norm Childers**, New Jersey and Florida. Fruit research and teaching.
1982	**Jim Moore**, Arkansas. Peach, blackberry and grape breeding.		**Elwyn Meader**, New Hampshire. Development of cold-hardy fruits.
	Roger Way, New York. Apple, cherry and elderberry breeding.	1992	**K.O. Lapins**, Agriculture Canada, British Columbia. Cherry breeding.
1983	**Gene Galletta**, USDA, Maryland. Strawberry breeding.	1993	**Wayne B. Sherman**, University of Florida. Peach and nectarine breeding.
	Don Fisher, British Columbia. Fruit researcher, administrator and author.	1994	**James Cummins**, New York. Apple rootstock breeding.
1984	**Victor Prince**, USDA, Georgia. Peach breeding.		**Elizabeth Keep**, East Malling, England. *Rubus* breeding and genetics.
	W.D. Armstrong, Kentucky. Extension specialist. Outstanding fruit extension work.	1995	**Floyd Zaiger**, Zaiger Genetics, Modesto, California. Stone fruit breeding.
1985	**Roy Rom**, Arkansas. Rootstock research.	1996	**Jules Janick**, Purdue University. Fruit breeding.
	Roy Simons, Illinois. Fruit anatomy.		**R.E.C. Layne**, Agriculture Canada, Ontario, Canada. Stone fruit breeding.
1986	**Hugh Daubeny**, Agriculture Canada, British Columbia. Small fruit breeding.	1997	**Dr. Arlen Draper**, USDA, Maryland. Small fruit breeding.
	Wallace Heuser, Michigan. Nurseryman. Fruit variety development.		**Dr. Derek Jennings**, Scottish Horticulture Research Institute, Scotland, U.K. Small fruit breeding.
1987	**No recipients**.	1998	**James R. Ballington**, North Carolina State University. Breeding Vaccinium, Rubus, and Fragaria.
1988	**Francis J. (Whitey) Lawrence**, Oregon. Strawberry and raspberry breeding.		
	Robert E. Lamb, New York. Disease-resistant apple breeding.		
	Grady Auvil, Washington. Tree fruit grower.		
1989	**Katherine Bailey**, New Jersey. Peach, nectarine, and apple breeding.		
1990	**Don Craig**, Nova Scotia, Canada. Strawberry breeding,		

Index

American Pomological Society, 150 Years ix
Apple
    'Cortland' 175
    'Delicious' 1
    'Empire' 94
    'Fuji' 137
    'Gala' 65
    'Granny Smith' 111
    'Golden Delicious' 19
    'Jonagold' 67
    'Jonathan' 25
    'McIntosh' 57
    'Wealthy' 83
    'York Imperial' 36
Baugher, T.A. 19
Blackberry, 'Marion' 183
Blizzard, S. 19
Blueberry
    'Bluecrop' 55
    'Sharpblue' 86
Brown, S.K. 17, 67, 175
Cactus Pear, 'Reyna' ('Alfajayucan') 116
Cahoon, G.A. 9, 152
Chandler, C.K. 5
Cherry
    'Bing' 60
    'Montmorency' 30
    'Napoleon' 17

Cranberry, 'Searles' 12
Crassweller, R.M. ix
Currant, Black, 'Crandall' 146
Daubeny, H.A. 38, 78, 90, 102, 133
Domoto, P. 1
Derkacz, M. 94
Draper, A. 55
Elfving, D.C. 94
Fan, X. 137
Fear, C.D 1
Finn, C. 183
Forshey, C.G. 94
Galletta, G.J. 72
Gmitter, F.G., Jr. 97
Gonzalez, S.P. 116
Grape
    'Concord' 9
    'French Hybrid' 152
Grapefruit, 'Marsh' 97
Hancock, J. 55
Hazelnut, 'Barcelona' 41
Hummer, K.E. 146
Iezzoni, A.F. 15, 30
Jacobo, C.M. 116
Jamieson, A.R. 149
Jennings, D. 125
Kiwifruit 22
Lawrence, F.J. 38, 102, 183

A History of Fruit Varieties   195

Lightner, G.L. 47
Lingonberry..106
Lyrene, P.M. 81, 86
Maas, J.L. 72
Maggs, D.H. 63
Maloney, K. 78
McGregor, G.R. 38, 78
Mehlenbacher, S.A. 41
Miller, A.N. 41
Moore, P.P. 90, 102
Myers, S.C. 47, 75
Nectarine, 'Sunred' 188
Okie, W.R. 47, 75
Orange, 'Valencia' 7
Patterson, M. 137
Peach
    'Elberta'..47
    'Flordaprince'...81
    'Redhaven' 15
    'Springcrest' 75
Pear, 'Bartlett' 5
Pecan
    'Desirable'...169
    'Schley' 119
    'Stuart' 33
    'Sumner' 142
    'Western Schley' 128
Pistachio, 'Sirora' 63
Pluta, S. 146

Proctor, J.T.A. 57
Raspberry
    'Glen Moy' 125
    'Heritage' 78
    'Meeker' 90
    'Willamette' 38
Rollins, H.A., Jr. 36
Rom, C.R. 60
Rom, R.C. 25
Roper, T.R. 60, 83
Sherman, W.B. 81, 86, 188
Soost, R.K. 7
Sparks, D. 119, 128, 142, 169
Stang, E.J. 12, 106
Strawberry
    'Earliglow'..72
    'Kent' 149
    'Senga Sengana' 133
    'Totem' 102
Strik, B. 183
Thompson, T.E. 33
Warrington, I.J. 22, 111
Way, R.D. 67, 175
Werner, D. 191
White, A.G. 65
Wilder Awards 191
Williamson, J.G. 188
Yoshida, Y. 137
Zurawicz, E. 133